Voices for Food Security

In 2012, leaders on every food-growing continent warned of future wars over food or the water to grow food.

"Accelerating climate change impacts will cause more than 100 million deaths and knock off more than 3% of GDP by 2030." **DARA** and the Climate Vulnerability Monitor, 2010.

"The threat from climate changes is serious, it is urgent, and it is growing, as more frequent droughts and crop failures breed hunger and conflict." – U.S. President and Nobel laureate **Barack Obama**'s U.N. Speech on Climate Change, 2009.[1]

"We know that a peaceful world cannot long exist, one-third rich and two-thirds hungry." — **Jimmy Carter**

"The war against hunger is truly mankind's war of liberation." – **John F. Kennedy**

"If you desire peace, cultivate justice, but at the same time cultivate the fields to produce more bread; otherwise there will be no peace." – **Norman Borlaug**

"The U.S. intelligence is preparing for the threat of a global war for water, which they believe is likely to occur by 2030." – Global Water Security, **U.S. intelligence Assessment**, 2012.

"For all the anxiety about the scarcity of oil, gas and vital minerals, the fiercest fight in the coming decades will involve food and the land it grows on." – **Michael T. Klare**, The Race for What's Left: The Global scramble for the World's Last Resources, 2012.[2]

"Man seems to insist on ignoring the lessons available from history." – **Norman Borlaug**

"Everybody thinks that the future is going to see fights over energy, it's far more likely to be primarily over food." – **Thomas Barnett**, The future of Fifth Generation Warfare: Follow the food!

"Over the next decade, water problems will create instability and state failures in countries important to the U.S. national security interests." Global Water Security, **U.S. intelligence Assessment**, 2012.

Peace Microfarms

A Green Algae Strategy to prevent War

Mark R. Edwards

Robert Henrikson

Make food, not war.

Let's save 100 million lives.

The Green Algae Strategy Series

AlgaeAlliance.com

AlgaeCompetition.com

Key words.

Food	Biofertilizer	Sustainability	Fertilizer
Water	Nutrient recovery	Ecosystems	Hunger
Regenerative	Organic farming	Smartcultures	Poverty
Agriculture	Climate change	Energy	Drought
Aquifers	Micronutrients	Environment	Soil
Soil crust	Global awareness	Algaculture	Malnutrition
Genetics	Renewable energy	Biotechnology	Pollution
Microalgae	Industrial farming	Ecology	Algae

ISBN- 978-1480141209
ISBN- 1480141208
Science / Biotechnology

Dedication

To Ian Parkinson for his dedicated service to his country.
A bomb blast in Afghanistan destroyed Ian's legs.
Ian has a great heart and mind that he wants to dedicate
to making the world a better place – without war.

Contents

A Green Algae Strategy to Prevent War

Human societies face famine, pestilence and community starvation.
Global climate chaos will destroy food crops and food stores.
Industrial agriculture will devour vital non-renewable resources.
Agricultural wastes will pollute fields, streams and ecosystems.

Our children will be left without the natural resources to grow food.
Our legacy will be depleted soil, dry aquifers and empty mines.
What will our children do for food?
History predicts war, unless we change our growing strategy.

To avoid war over our floundering and unreliable food supply,
Growers need independence from weather and fossil resources.
Peace microfarms offer a novel solution with microcrops.

Peace microfarms cultivate freedom foods that are 30 times more
productive than industrial agriculture. Growers practice abundance as
they nurture healthier foods for people and our planet that:

- Grow without cropland.
- Are naturally biodiverse.
- Thrive with no fresh water.
- Use no or minimal fossil fuels.
- Consume no or few inorganic fertilizers.
- Use no or minimal pesticides or poisons.
- Recover, recycle and clean waste streams.

These climate independent foods allow growers to produce food all
year round independent of altitude, latitude, geography or politics.

Our challenge: R3D, research, develop, demonstrate and diffuse
Peace microfarms globally to enable people to produce food locally.

Let's leave a superb legacy for our children,
Healthier food, sustainable for many generations,
With plentiful natural resources and clean, green ecosystems.

Forward

Can peace microfarms prevent war? Yes, with your help.

Hundreds of books chronicle war and the terrible hunger that precedes and follows war. No prior books consider microcrops such as algae solutions for food security, water security or national security. Peace microfarms offer novel sustainable solutions to prevent wars over food. Microfarms may provide the only means to avoid conflict over the finite natural resources needed to produce modern foods.

Microfarms are currently too complex and expensive for widespread adoption. Our challenge is to design microfarms that are easy to deploy and operate at minimal total cost. We need to engage a broad community to deliver sustainable and affordable microfarms to the many people who want to grow food for their family and community.

We believe this peace microfarm strategy can be implemented within three years for less than the cost of one day of modern war.

Please engage in the fascinating global open source collaboratory at AlgaeCompetition.com. Your engagement could make the difference as we work to:

Replace firearms with microfarms.

Key Terms

Fossil foods are unsustainable commercial foods that escalate with fuel prices. Industrial foods suffer from hidden hunger, deliver empty calories, fail in bad weather, are dependent on costly non-renewable resources and create massive waste and pollution. Industrial foods are unsustainable since they rely on fossil resources.

Freedom foods redesign our food supply from the foundation of the food chain with microcrops. Freedom foods free consumers for smart food choices, growers from fossil resource cost and consumption and ecosystems of waste and pollution.

Peace microfarms liberate growers from dependence on increasingly expensive cropland, fresh water and other non-renewable resources. Peace microfarms avoid conflicts over diminishing natural resources by growing microcrops using abundance methods. Microcrops include the full spectrum of microorganisms such as algae, yeast, fungi, bacteria, archaea, plankton and many others. Microcrops deliver sustainable advantages over field crops.

Peace microfarms produce 20 to 30 times more food per acre every year than modern agriculture. Microfarms scale to any size and may be sited practically anywhere, including cities. Growers recover low cost nutrients from sterilized waste streams and transform them into valuable freedom foods and other products. Growers use abundance methods to assure a sustainable food supply for many generations.

Abundance methods offer an alternative for growing food and other forms of energy with plentiful resources that are cheap and will not run out – primarily solar energy, CO_2 and sterilized waste stream nutrients. Abundance allows affordable organic food production in practically any climate, altitude, latitude or geography.

Nutralence is the food nutrient availability and density in food. Freedom foods offer 200% more nutrient availability and 200 to 500% more nutrient density per bite than industrial foods. Foods with natural nutralence do not deliver the empty calories that cause obesity and diabetes.

Chapter 1. Food Security and War

*I am here to sound the alarm about our direction as a
human family, especially global warming and rising food
prices.*   *– **Ban Ki-moon**, U.N. Secretary General, 2012.*

*More than 100 million people will die and global
economic growth will be cut by 3.2% of GDP by 2030 if
the world fails to tackle climate change.* – The Climate
Vulnerable Forum, **DARA**, a partnership of 20 countries.[3]

Affordable food represents the primary threat to world security. Peace microfarms offer a new strategy that may save 100 million lives by eliminating the need to fight over cropland, freshwater or other finite resources needed to sustain our food supply.

When people lack food for their families, they take action that escalates from demonstrations to riots to revolution. Once wars ignite, conflicts consume resources at an astonishing rate and often leave both sides with less food, money and means to produce food.

Leaders in practically every country predict war over food. Scientists, economists, political leaders and authors also forecast conflicts over increasingly scarce food and the finite resources required by industrial agriculture – cropland, water, fossil fuels, fertilizer and chemicals.

Disruption from weather or to any of the many resources necessary to grow crops, or the supply chains that distribute food, will put millions of people in in peril. Disruptions become increasingly likely with bad weather, water scarcity and escalating fuel and fertilizer costs.

War becomes more likely as populations expand and new consumers eat higher on the food chain. Farmers have no idle land. Food stocks are the lowest in history. Countries are living harvest to harvest. We have more hungry people on earth than ever before.[4]

State of hunger

The 2010 FAO *State of the World* estimates 925 million hungry people in the world. This represents 13%, or almost 1 in 7 people are hungry. Hungry people get less than 2,000 calories a day. The number understates the larger number of people who are malnourished due to the lack of quality food or who suffer from nutrient deficiencies. Roughly half the people on Earth are food insecure, without reliable access to good food.

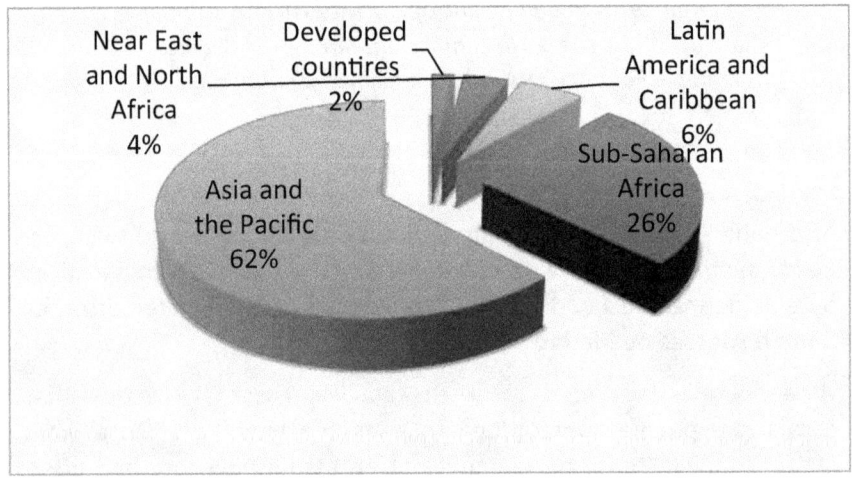

Figure 1.1. *Hungry People in 2010 = 925 million*

Each year the world adds nearly 80 million people. Every day there are 219,000 hungry new people at the supper table. Many new people will have empty plates. Many families are forced to experience foodless days because they cannot afford food. A recent survey by Save the Children reports that the 24% of families in India have foodless days, Nigeria 27% and Peru 14%.[5] Millions of children are dangerously hungry, some too weak to walk to school. The FAO reports that 50% of children globally have nutrient deficiencies and many are physically and mentally stunted.

In the U.S., blessed with ideal growing conditions, one out of five Americans – over 60 million citizens – receives food support because they are food insecure and hungry.[6] Food support includes the 48 million people on food stamps that provide about $1.50 per meal. Dollar fifty meals eliminate the opportunity for families and their children to participate in the American dream. Poor Americans lack food freedom. They must feed their children cheap, nutrient deficient foods that cause obesity, fatigue, dull brains and diabetes.

One out of every four American children struggles to get enough food for their bodies and minds to develop properly.[7] Food insecurity forces our children to eat cheap food, which typically delivers loads of fat, salt and cholesterol, but poor taste and color. These foods deliver empty calories, largely devoid of nutrition. Malnourished children incur developmental impairments that limit their physical, intellectual and emotional development.[8] Developmentally impaired children place a huge drag on families, medical facilities and education. At a time when nearly half the global population is food insecure, demand for food removes more grain from food stores.

New consumers

Today, two out of three world citizens subsist on local or imported grains and local plants because they cannot afford meat. Livestock are the single largest user of land as meat production accounts for 70% of all agricultural land and 33% of the land surface of the planet.[9] In order to produce meat, roughly one-third of the world's food grains go to feed livestock. In addition, livestock are responsible for 18% of all greenhouse gases, which is more than all cars and SUVs combined.[10]

Livestock contribute 37% of the methane and 65% of the nitrous oxide to the atmosphere, which are 20 times worse greenhouse gases than CO_2. A single cow—calf pair produces more gas emissions than a person driving 8,000 miles in a midsize car.[11] Meat is dramatically underpriced relative to plant foods because meat production benefits from both crop and Big Oil subsidies. Meat prices reflect no environmental accounting even though meat producers extract trillions of gallons of fossil water, billions of gallons of fuels and millions of tons of chemicals. In addition, agricultural waste streams are the primary polluters of surface and groundwater yet the public costs and resource losses are absent from meat prices.

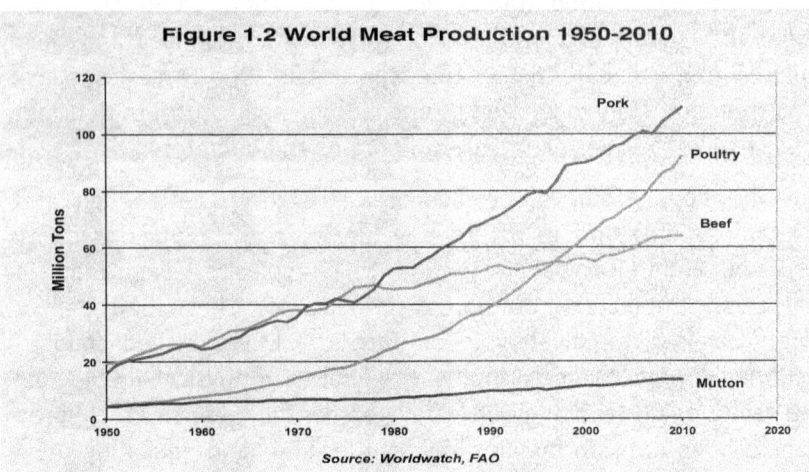

Figure 1.2 World Meat Production 1950-2010

Source: Worldwatch, FAO

New consumers have become carnivores and demand animal-based foods, which create an ecological cost of multiple pounds of grain for each pound of meat or dairy. Americans continue to increase per capita meat consumption and people around the world desire to emulate the consumption patterns of the rich. Feed conversion ratios, grain to meat, are:

- Farm raised tilapia or catfish   2:1
  (most fish weight gain is water trapped in tissues)
- Poultry – chicken or turkey   3:1
- Beef, pork or lamb   8:1

Since one ton of grain consumes 1,000 tons of water, one ton of beef on the hoof uses 8,000 tons of water. However, beef on the hoof contains 57% waste – bone, hide, fat, stomach and unsellable organs. A ton of beef on the hoof yields only 43% sellable beef, so 2.3 tons of beef on the hoof are necessary to yield a ton of sellable beef.[12] Therefore, a ton of sellable beef, based on grain input, consumes 18,400 tons of water or 9.2 tons of water per pound of beef.

Actual water required by beef is even higher because feed conversion ratios reflect only the grain and ignore the cow's own water consumption. A 1200-pound cow drinks about 36 gallons of water a day.[13] The cow's thirst adds another 20% to the water cost of beef. A pound of sellable beef consumes 11 tons of water or 2,650 gallons. In spite of the huge water cost of meat, meat production continues to increase due to demand, especially in India and China.

The water cost of sellable beef translates to a water cost for a quarter-pound hamburger of 3,000 gallons plus another 400 gallons for the cheese, 100 gallons for the lettuce and tomato and another 100 gallons for the bun. A gallon of milk costs 4000 gallons and a pound of coffee consumes about 2,650 gallons of water.[14]

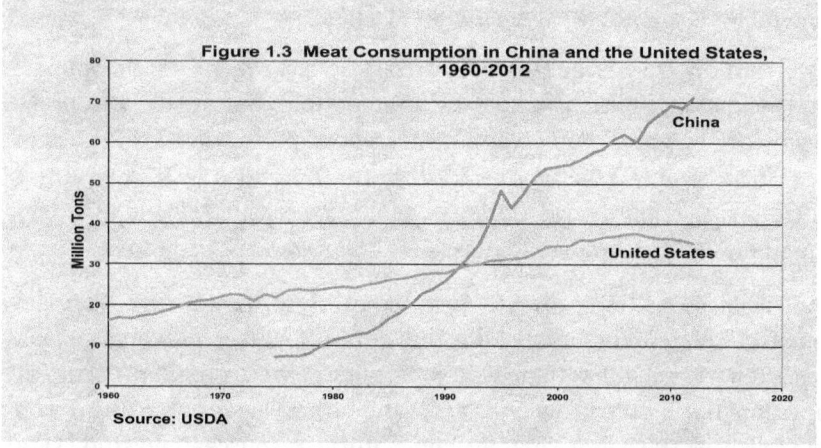

Figure 1.3 Meat Consumption in China and the United States, 1960-2012

Source: USDA

China has twice the number of new consumers as the U.S., 600 million, who demand higher value foods that create a high water exchange. China leads the world in production and consumption of meat. Chinese per capita meat and milk consumption have doubled in the last 20 years.[15] China's increasingly high demand for grains combined with its distressed ecology and water problems (the two largest rivers, the Yellow and Juma, have gone dry, as have nearly all the lakes around Beijing) will dramatically reduce world grain supplies. When drought, storm or pest invasion destroys food production in a major food-growing region such as China, India, Russia, Ukraine, Argentina or the U.S., food markets will be destabilized worldwide.

The Stockholm International Water Institute's *Feeding a Thirsty World* addresses water scarcity and food production with the prediction that most diets by 2050 will be 95% vegetarian because there is simply not enough water to support meat animals.

Food supply vulnerability

Even with all the advances in technology and weather forecasting, modern food production has become more vulnerable to small climate disruptions. When past civilizations experienced drought, such as the U.S. Dust Bowl of the 1930s, people could simply migrate to another area with more water. Current populations are far denser and millions of people have nowhere to flee.

Interruption from weather or terrorist strike to any of the mines, stores or distribution networks that supply the many resources necessary to grow crops will put millions of people in risk. Breaks in the long supply chains that distribute food also will put entire societies in jeopardy. Disruptions become increasingly likely with bad weather, water scarcity and massively pollutive waste streams.

Industrial crops' long growth cycle creates substantial risk. Farmers must cultivate most crops for the entire 120-day growing season before harvest. If weather, water, fertility or pest problem occurs at any time during the growing season, the farmer loses not just the crop but also all the labor, money and natural resources invested. Late season losses mean the farmer also misses the opportunity to grow a short-season crop to feed the family.

The 2012-year demonstrates how vulnerable industrial agriculture makes societies with small changes in weather. A 2.5 F increase in temperatures resulted in huge grain and other food crop losses. Crop insurance losses are estimated at $20 billion.[16]

The structure, reliability and sustainability of our food supply should be a top priority for national defense, security, health and education. Politicians and policy leaders ignore the strategic value of our food supply – to our peril. The U.S. budgets tell the story. National defense receives nearly a trillion dollars annually but produces no food.

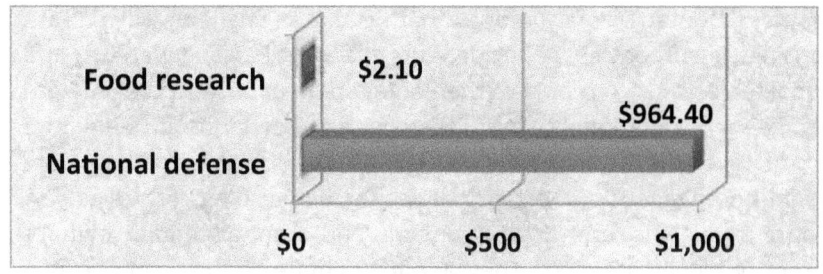

U.S. 2012 Budget in Billions

Research on alternatives to industrial agriculture has been trivial. Organic production receives less than 2% of the USDA R&D budget. Investments in new methods, such as abundance, round to zero.

Food wars

Food wars are caused by crop failure that result from climate change and diminishing non-renewable resources required for industrial food production, Figure 1.4. Resources become extinct when they are no longer affordable or available locally.

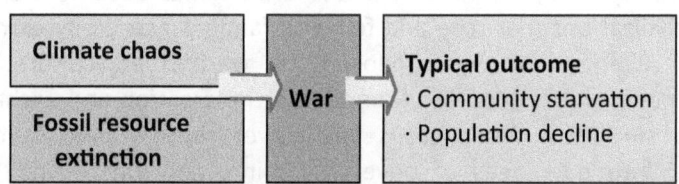

Figure 1.4. Causes of Food Wars

Climate chaos

Climate chaos brings a wide range of threats to food security. Robert Kaplan envisioned in 1994 the core foreign-policy challenge for the 21st century as the "Political and strategic impact of surging populations, spreading disease, deforestation and soil erosion, water depletion, air pollution, and rising sea levels." He predicted that these events would prompt mass migration and, in turn, incite group conflicts.[17] A 2007 report from Christian Aid estimated 1 billion people would be forced to leave their homes by 2050, which could destabilize whole regions where increasingly desperate populations compete for dwindling food and water.[18]

Peace Microfarms

Thomas Homer-Dixon argues that climate change will create uprisings, insurgencies, genocide, guerrilla attacks, gang warfare, and global terrorism.[19] A report to the Pentagon on implications of climate change for U.S. national security sketches epic scenarios, including the risk of reverting to a Hobbesian state of nature whereby humanity would be engaged in constant battles for diminishing resources.[20] A report by eleven retired U.S. generals and admirals added military authority to climate risks. They argue, "Climate change can act as a threat multiplier for instability in some of the most volatile regions of the world and that this presents significant national security challenges for the U.S."[21]

Bad weather for crops causes price spikes, fear, food riots and war. The French Revolution in 1784 ignited due to escalating food prices. "Qu'ils mangent de la brioche," (Let them eat cake) uttered by Marie Antoinette, was in response to the poor demonstrating for bread to eat. Crops failed due to extreme weather from El Niño, amplified by the 1783 volcanic activity at Laki and Grímsvötn, Iceland.

In *Empires of Food*, Evan Fraser and Andrew Rimas document over a dozen civilizations that rose and fell with famines caused by extreme weather events that were followed by war. Cities, culture, art, government, and religion are founded on the creation and exchange of food surpluses. When crops fail, power centers shift. Cultures descend into dark ages of poverty, famine, and war – and then starvation.

Food causes war and plays a central role in war. Barbara Clark Smith in *Food Rioters and the American Revolution* chronicled how scarce food, notably sugar, tea and bread in 1775, ignited the American Revolution. Historian Lizzie Collingham reported in *The Taste of War: WW II and the Battle for Food* that 20 million people, about half prisoners, died of starvation in World War II.

The 2011 Arab Spring parallels the revolutions across Europe in 1848. The "Spring of Nations" experienced a year food riots the escalated to revolutions throughout Europe. The "hungry '40s" saw a decade bad weather that caused failed harvests. Hungry people in Europe became angry. Angry people organize to bring down governments.

8

Climate Chaos Impacts	Burning food supplies is a classic war strategy intended to undermine sustenance and destroy the will to fight. Burning crops was a common action during the American Revolution and the Indian Wars. During the American Civil War, General Sherman burned a 50-mile swath from Atlanta to Savanna on sea to drain the will to fight from the Confederacy.

Temperature
- Heat spikes
- Cold spikes
- Hot dry winds
- Heat stress
- Earlier spring
- Later fall
- Unpredictability

Water scarcity
- Drought
- Faster evaporation
- More irrigation
- Stress on aquifers
- More water conflicts

Atmosphere
- Higher temperatures
- Higher air moisture
- More fierce storms
- New rain patterns
- Massive blowing dust
- Accelerated erosion

Land surface
- Hotter temperatures
- Loss of snow pack
- Loss of glaciers
- Rivers go dry
- Amplified wildfires

Oceans
- Higher surface temps
- Rising sea levels
- Higher ocean acidity
- Loss of sea ice

Burning food supplies is a classic war strategy intended to undermine sustenance and destroy the will to fight. Burning crops was a common action during the American Revolution and the Indian Wars. During the American Civil War, General Sherman burned a 50-mile swath from Atlanta to Savanna on sea to drain the will to fight from the Confederacy.

In *Starving the South*, Andrew Smith argues the naval blockade that resulted in starvation and hunger across the South was more important than the musket balls fired in winning the Civil War.

Christian Parenti in the *Tropic of Chaos: Climate Change and the New Geography of Violence* examined how societies such as Somalia, Afghanistan, Brazil and Mexico have responded to the many unpredictable vulgarities of climate change. He concludes that the risk of war rises with scarcity.[22]

Climate chaos stresses countries and these changes threaten American national security. Christian Parenti provides context for the choice of Afghanistan farmers to grow opium poppies. Afghanistan has suffered a prolonged drought and poppy uses only 20% of the water wheat requires. The U.S. military burns the farmers' crops, which creates animosity.

Taliban supports the farmers and helps them feed their children. Without food or money to buy food, farmers and their communities would face starvation.

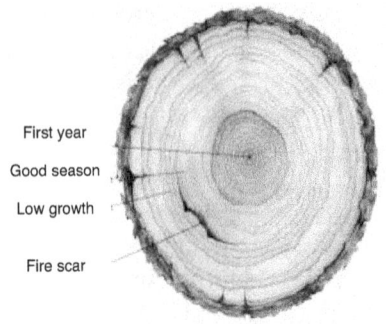

First year
Good season
Low growth
Fire scar

Tree rings record poor growth years, that link to crop failure and war.

A team led by David Zhang from the University of Hong Kong analyzed tree rings to estimate the temperature swings in China over the last thousand years. They reviewed 899 wars fought in China between 1000 and 1911 and found a strong correlation between the frequency of warfare and temperature changes.[23]

Records showed that weather changes resulted in food price inflation, followed by war, famine and population decline. The findings suggest that worldwide price cycles in recent centuries have been driven mainly by climate change.

A second team led by David Zhang examined temperature data and climate-driven economic variables during the "golden" and "dark" ages in Europe and the Northern Hemisphere during the past millennium. Their findings align with the China studies. The data indicated that climate change was the ultimate cause, and climate-driven economic downturn was the direct cause, of large-scale human crises in preindustrial Europe and the Northern Hemisphere.[24]

Farmers used organic production during the many wars in China. Some Chinese farmers continue the practice of using "night soil," human excrement on their fields. Organic production is more resilient to weather effects than industrial agriculture but yields still drop or fail in bad weather years.

Population decline following war occurs from community starvation, which must be the most agonizing way for humans to die. Families are forced to watch helplessly while their weakest suffer. They see their children and elders in prolonged and excruciating misery while the human body consumes its own tissues in a desperate attempt to sustain energy. The victim's skin changes color and loses elasticity while the stomach loses its ability to digest. Eyes sink in their sockets and victims lose their memory and become weak and disoriented. Death usually comes from diseases such as dysentery, pneumonia or heart failure that mercilessly attack the weakened body.

Struggles over food shortages are occurring with increasing frequency. Over 40 countries experienced severe food riots in 2008 when food prices spiked. Recent food insurrections disrupted national economies, brought down governments, spurred food theft and resulted in thousands of deaths. Countries created policies that prohibited food hording, waste and food exports.

Rising food prices in 2011 ignited the Arab Spring, with revolutions across North Africa. People may tolerate despots as long as they provide sufficient affordable food. Revolutions spurred by food scarcity not only kill or maim many able bodied young men but also destroys the infrastructure for growing and transporting food.

Stanford University professor David Lobell models climate change and concludes, "The impact of climate change on food production can already be seen, and will worsen as climate change gathers pace." Lester Brown repeats his mantra in *Full Planet, Empty Plates: The New Geopolitics of Food Scarcity*, "The world is only one poor harvest away from disaster." While climate chaos threatens food security, the depletion of finite resources amplifies threats to the food supply.

Fossil resource extinction

Countries go to war when the vital resources needed to produce food become extinct locally and market prices are too high, Table 2.1. Most experts predict war over the obvious resources, cropland, water and fuels. Careful analysis shows wars may be fought over macronutrients such as phosphorus and micronutrients such as copper and zinc, which are in very limited supply. Failing sufficient zinc in the soil, crops fail to fruit, (grow seeds), which eliminates not just the food portion of the crop but subsequent years' seeds.

11

Primary fossil resources	Micronutrients	
1. Fertile soil	12. Oxygen	(O)
2. Freshwater	13. Hydrogen	(H)
3. Fossil fuels	14. Sulfur	(S)
4. Inorganic fertilizers	15. Magnesium	(Mg)
5. Pesticides and herbicides	16. Boron	(B)
6. GE seeds	17. Copper	(Cu)
Macronutrients	18. Chlorine	(Cl)
	19. Iron	(Fe)
7. Nitrogen (N)	20. Molybdenum	(Mo)
8. Phosphorus (P)	21. Manganese	(Mn)
9. Potassium (K)	22. Nickel	(Ni)
10. Calcium (Ca)	23. Zinc	(Zn)
11. Carbon (C)		

Table 2.1. Fossil resources required for industrial agriculture
Note: GE seeds are a finite resource because these seeds
require intensive fossil resources to produce.

The FAO reports that 33% of the world's cropland has become so degraded it has been abandoned in the last 40 years and continue at 29 million acres a year.[25] The U.S. net cropland losses from 1982 to 1992 covered an area the size of New Jersey.[26] About two million hectares, (five million acres) of fertile agricultural lands are lost to production every year due to industrial agricultural methods that:

- Wear out the soil by extracting the nutrients and humus.
- Erode the soil by cultivation makes it vulnerable to erosion.
- Increase soil salinity from fertilizer or irrigation salts.[27]

When fields become unfertile, they must be abandoned because the soil does not support crops. Half the remaining cropland globally is so degraded it takes twice as much fertilizer and three times more irrigation water to achieve normal crop yields. Nearly all – 94% – of Iran's agricultural land is degraded from erosion or salination.

Nature requires about 500 years to replace 25 millimeters (1 inch) of lost topsoil. The minimal soil depth for agricultural production is 150 millimeters (5.9 inches) but many crops need deeper soils. Fertile soil is a nonrenewable, endangered ecosystem that with degradation systemically diminishes crop yields until the soil become unfertile.

The Environmental Working Group collaborating with the USDA found that the rich, dark soil in America's Heartland is being swept away at rates many times higher than official estimates. In some places in Iowa, recent storms have triggered soil losses that were 12 times greater than the USDA average for the state. A single storm can strip 64 tons of soil per acre from cropland, according to researchers using the new measurement techniques.[28]

Land grabbing

The current popular solution for cropland scarcity, land grabbing, is neither ethical nor sustainable. China, Saudi Arabia, South Korea, and United Arab Emirates have bought huge tracks of land to produce food for export back home. Incredibly, the croplands are in countries that receive emergency food aid. Land leases run for 25 to 99 years, often for less than $1 per acre per year.

Independent agencies have charged politicians with corruption in the land leases. Farmers and indigenous people that have lived on those lands for generations often find out about deals only as they are forced from their land. In 2012, more than one million Ethiopians forcibly relocated by their government.

When the displaced indigenous people face community starvation, no level of military troops will be able to stop them from taking crops. Land grabbers will find their investments foiled by precisely the people that, in their greed, they displaced.

Resource consumption

Modern industrial agriculture improved human societies but its Achilles' heel is that food production depends on massive consumption of fossil resources – fertile soils, freshwater delivered at just the right time, fossil fuels, inorganic fertilizers and agricultural chemicals and poisons. When just one resource became extinct locally, entire civilizations perished. History will repeat wars and community starvation as regions and countries run out of critical inputs for modern food production.

Elizabeth Kolbert, in *Field Notes from a Catastrophe,* chronicled a list of sophisticated cultures that sustained themselves for hundreds of years and then crashed when their soils no longer supplied sufficient food, including:

- Tiwanaku, Lake Titicaca in the Andes – crash: A.D. 1100, drought
- Classic Mayan civilization – crash: A.D. 800, drought
- Old Kingdom of Egypt – crash: 2200 B.C., drought
- Akkadian empire – crash: 2200 B.C., drought[29]

Droughts amplify the need for irrigation. Irrigation systems designed for normal years are unlikely to deliver sufficient water in hot years when crops may need two or three times more water. Heat also intensifies soil salts. Irrigation water evaporation leaves a white crust of irrigation salts that diminish and eventually destroy soil fertility.

Food crops are voracious water consumers. A ton of grain consumes 1,000 tons of water of water. The U.S. is the largest exporter of wheat, corn and soy to the world. When the U.S. exports a ton of wheat, buyers are effectively buying 1,000 tons of water. When the U.S. exports a ton of beef, buyers are buying 22,000 tons of water.

Large sections of the Heartland's aquifer, the Ogallala, will run dry, probably about 2030. At about the same time, aquifers are predicted to crash in California, India, China and other countries. What will our children do for food?

At the time the world needs more food in 2012, the U.S. Midwest suffered its worst drought in 56 years. The USDA cut its corn forecast to 274 million tons, a six-year low. The heat and drought in Russia caused crop losses that is likely to result in a 20-year low. The loss of U.S. and Russia grains will push up prices on global markets.

These yield losses may spark more food riots and revolutions. The next section examines current agricultural research in search of models for expanded food production. The world desperately needs a solution to food security – soon.

Chapter 2. Food Security Solutions

Extreme weather events in a single year could bring about food price spikes of comparable magnitude to two decades of projected long-run price increases. **Oxfam** Issue Briefing, *Extreme Weather, Extreme Prices, The Costs of Feeding a warming World, September 2012.*

Resource depletion is accelerating even faster than predicted and the competition for scarce resources has intensified. If we do not adapt, it will likely end in war and massive starvation. **Michael T. Klare** *in The Race for What's Left: The Global Scramble for the World's Last Resources, 2012.*

Severe climate events, natural disasters and resource depletion recently underscore the need for immediate action. New records were set in 2012 for searing heat and drought, the number and severity of wildfires, extreme storms, floods, West Nile deaths, sea ice melt and crop insurance loss. The USDA declared over half of all counties in the U.S. natural disaster areas. Heat and drought engulfed 90% of corn and soy acreage.

Insured losses of $116 billion in 2011 were the industry's second costliest for natural disasters in a half a century. A devastating earthquake hit Japan and severe flooding destroyed crops and infrastructure in Thailand, China, India and Pakistan. Wildfires destroyed huge tracks of land in Russia, China and the U.S. Total losses in 2012 could be worse due to more severe storms such as Superstorm Sandy, flooding, storm surge, heat, drought and wildfires.

Food security actions fall into three groups, political rhetoric, technical advances and genetic engineering, Figure 2.1.

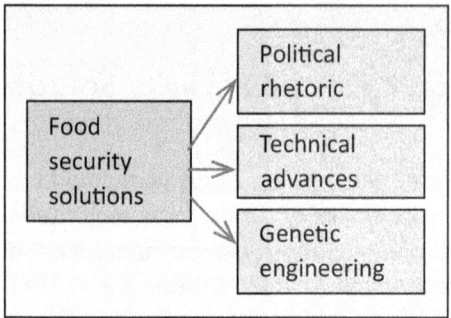

Figure 2.1. *Food security solutions*

Leaders speak

A parade of over 50 highly credible voices call for a doubling of the world's food supply in the next 30 years, including:

- Robert Zoellick, World Bank president[30]
- Ban Ki-moon, United Nations Secretary General[31]
- John Beddington, the United Kingdom's Chief Scientific Adviser[32]
- LaMar Lemmons, Michigan State House of Representatives[33]
- Hugh Grant, Chairman, President and CEO of Monsanto[34]

Political statements may improve the standing of politicians and leaders but produce no food. Recommendations to increase food production would provide value, if countries had the political will to take action. A few countries, such as China and India are making strong commitments to sustainable solutions, especially green energy.

Many world leaders have lost touch with agriculture. The proof is their lack of investment in agricultural innovation. Leaders may be unaware that peak food may have already have occurred.[35] World food production may not have the capacity to expand at all or even to sustain current production with conventional agriculture.[36]

Economists and agricultural economists fail to factor into their business models the cost of agriculture's self-destruction, over-consumption of finite resources and climate shocks. Nicholas Stern, former chief economist at the World Bank, examined the cost of failing to incorporate the climate change costs and burning fossil fuels and concluded the cost would be in the trillions of dollars.[37] Estimates for the cost of environmental pollution agricultural pollution in the U.S. exceed $40 billion annually.

Unfortunately, political statements are unlikely to motivate sufficient food production. Many pin their hopes on expected improvements in agriculture technologies. Technical innovations require investments and few countries are investing in new agricultural technologies.

Technical solutions

The set of 15 research areas Worldwatch assembled for Earth Day 2012 summarize active lines of research to improve agriculture and food security. Research initiatives tend cover more social issues than crop yield and productivity.

Guaranteeing the Right to Food. Harnessing the Nutritional and Economic Potential of Vegetables. Reducing Food Waste. Feeding Cities. Getting More Crop per Drop. Using Farmers' Knowledge in R&D. Improving Soil Fertility. Safeguarding Local Food. Biodiversity.	Coping with Climate Change. Harnessing the Knowledge and Skills of Women Farmers. Investing in Africa's Land. A New Path to Eliminate Hunger. Improving Food Production from Livestock. Going beyond Production. Moving Ecoagriculture into the Mainstream.

Worldwatch Nourishing the Planet project highlights

Industrial agriculture uses a set of mechanical technologies that have improved crop productivity, including water and fertilizer application. Incremental improvements increased crop productivity from 1960 to 1980, but per capita yields have slowed substantially in recent years, Figure 2.1. Total grain production increases also have slowed since 1980. Europe has hit a glass ceiling. Rice production has not increased in Japan for 17 years. South Korea and China's has plateaued and may not increase.

The outlook is not good. Each 1°C (2.5 °F) temperature rise above the optimum during the growing season cuts wheat, rice, and corn yields by 10%. A 3°C rise cuts yields by 50% and a 5° C rise destroys the crop.

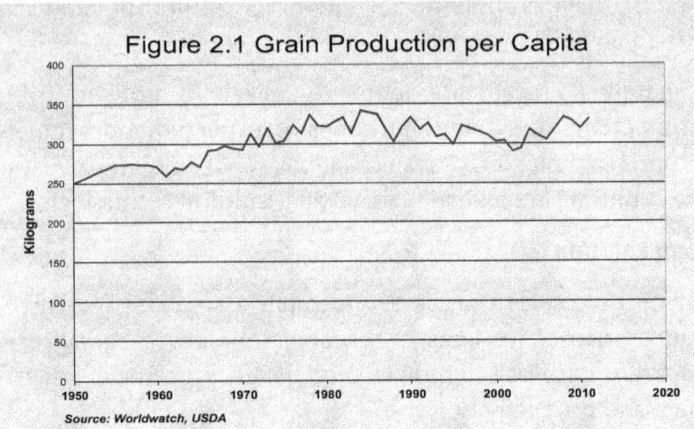

Figure 2.1 Grain Production per Capita

Source: Worldwatch, USDA

New technologies are needed to increase water productivity because over 40% of the world's grain is grown with irrigation. India, China and the U.S. will face serve water shortages as major aquifers crash, probably around 2030, if not earlier.

Incremental improvements in water conservation focus on minimizing runoff with contour plowing, no-till and drip irrigation systems. Contour plowing is not new since the 1934 *Yearbook of Agriculture* reported that 35 million acres of formerly cultivated land in the U.S. had been destroyed for crop production and another 100 million acres had lost most of the topsoil. In good conditions, contour plowing provides a 5 to 10% yield improvement. Most U.S. farmers adopted the practice decades ago.

Many farmers have adopted no-till or conservation farming to improve fertility and avoid compaction, erosion and loss of organic matter. Tilling removes weeds and shapes the soil but accelerates erosion. No-till farmers sow seeds with minimal soil disturbance.

No-till reduces labor and cultivation costs but does not increase yields. No-till requires a combination of new techniques that slow adoption including equipment, pesticides, crop rotation, fertilization and irrigation adapted for local conditions. A USDA study showed that the first inch of no-till soil is up to seven times less vulnerable to erosion than plowed soil.

Irrigated croplands often lose over half the water from leaks or evaporation before it arrives to the field. Flood and sprinkler irrigation lose another 50% from evaporation before the water enters the soil. Lose or sandy soil is so porous that water quickly percolates below the root zone. As farmers expand fields, they often have to use sub-optimal soils that require additional irrigation.

Drip irrigation, is a 30-year-old technology that allows water to drip in the plant root zone, saving water and fertilizer. Farmers often use drip tape buried in the middle of the root zone. Unfortunately, some areas cannot use drip due to particulate or salt build up that clogs emitters.

Farmers are trying a variety of methods to reduce needed irrigation water because they must compete with cities for water. Ethanol subsidies work against water conservation. A corn crop in the western U.S. requires about 3 to 5 acre-feet per acre, about four million gallons of water. A typical household consumes about an acre-foot of water annually.

Each gallon of ethanol produced with irrigated corn consumes 3,000 gallons of freshwater for the corn feedstock and additional water for refining.[38] Some areas on the high plains get about 1/3 of their water as rain so their ethanol consumes 2,000 gallons of water per gallon of ethanol. The U.S. harvested about 130 million tons of corn as feedstock for 14.2 billion gallons of ethanol in 2011, which displaced about 3% of U.S. oil imports.[39] Ethanol producers received about $6 billion in direct subsidies and several billion more for water and power. Ethanol production not only removed 130 million tons of corn from food markets but also consumed over two trillion gallons of irrigation, largely from nonrenewable aquifers.

New technologies in fertilizer application are important because the cost of fertilizer has risen to 40% of the total growing cost for some farmers. The price of nitrogen, (N) fertilizer rises with oil prices because 90% of the N fertilizer price comes from the energy required to produce it. New side-dressing N applicators reduce fertilizer waste but are expensive and are useful only during the early growth phase when plants are short.

The price of one fertilizer, phosphorus, (P) rose by 700% over a recent 14-month period. There is no artificial substitute for P in agriculture and crops fail to grow without P. As world reserves of this critical natural resource diminish, prices will skyrocket. Scientists at Linköping University in Sweden and Arizona State University predict peak phosphorus will occur by about 2031.[40] The few U.S. phosphate mines will be empty within 20 years. Researchers at the Sustainable Phosphorus Initiative at Arizona State University are conducting research on algae's ability to recycle P from waste streams to fields.[41]

No new mechanical agricultural technologies currently offer more than modest yield improvements. No industrial agricultural research in soil science, irrigation or mechanical equipment promises anything close to a 10% yield improvement. Therefore, many pin their hopes on transgenic seeds.

Genetic engineering

Genetically engineered, (GE) seeds were introduced in 1996 with the promise that they would increase yields of healthy foods while reducing crop inputs. So far, the results for genetically modified foods, (GMO) are mixed. Farmers adopted GE seeds so fast that within 15 years, 92% of U.S. corn and soy come from GE seeds. The European Union strictly forbids GE foods and requires GE food labels. The California Right to Know campaign put GE labeling on the Nov 6, 2012 ballot but the initiative was defeated. The advertising campaign asked "Are Your Children Eating GMO Pesticide Sweet Corn?"

A series of studies have revealed that genetically engineered foods can pose serious risks to humans, wildlife and the environment. Human health effects include higher risks of toxicity, allergenicity, antibiotic resistance, immune-suppression and cancer. Environmental impacts include uncontrolled biological pollution threatening numerous microbial, plant and animal species with extinction. Transgenic seeds may contaminate heritage plants with novel and possibly hazardous genetic material.

A nasty secret of current GE seeds is that they increase vulnerability to climate change and accelerate fossil resource depletion. New GE seeds are more productive because engineers trick the plants into putting more energy into seeds and less into roots. Shorter root structures require substantially more finite crop inputs, water, cultivation, fuels, fertilizer and pesticides. Shorter roots also make the crops less able to tolerate stressors from heat, drought and pests. The GE seeds need more six times more expensive fertilizers and pesticides than traditional seeds.[42] On top of these costs, farmers have to buy new GE seeds every year, largely from Monsanto.

GE seeds reduce biodiversity. Over 90% of the world's food is derived from just 15 plant and 8 animal species.[43] The cost and time required obtaining certification for GE seeds results in a single cultivar for corn, soy and other seeds. Monocultures of identical crops put the entire food supply at risk from a single pest vector.

Industrial agriculture and GE seeds have not been kind to Indian farmers. Over 8 million Indian farmers quit farming during the last decade due to rising GE crop input costs that escalated farmer debt beyond their means. The GE seeds required a five times increase in irrigation, forcing farmers to dig 21 million new wells. Many wells ran dry within two years. Electricity blackouts are common in areas where over half of the electricity is used to pump water from depths of two thirds of a mile.[44] Blackouts on consecutive days in 2012 deprived 600 million people – half of India's population – of electricity and transportation.

India's Crime Records Bureau reported that since 1995, 250,000 farmers in India committed suicide because their farms could no longer provide for their families.[45] Suicide rates are increasing to one every 12 hours. Additional millions of farmers and family members have died or become disabled from agricultural chemicals and poisons. Trains from the city of Chotia Khurd in northern India are now called cancer trains because so many people from the farming villages must go to the city for cancer treatments.

Monsanto and a few other companies promise transgenic seeds that are more heat, drought and salt tolerant. Climate tolerant transgenic seeds are still a decade away and no one knows how well they will work. Several scientific organizations including the Environmental Working Group and the Union of Concerned Scientists are concerned about the effects of transgenic seeds on human and animal health and the impacts of those seeds on the natural environment.

Real solutions?

The current lines of agricultural research may improve crop yields in ways yet unknown. No one seems to be hinting at the mechanisms to substantially enhance the food supply. Expanding populations are going to need every available method of food production.

We propose a new solution – an alternative food supply where microcrops are grown in peace microfarms.

An alternative food supply?

What if farmers could grow healthier foods that overcome the major industrial agriculture deficiencies, including foods that grow free of:

- Climate, weather or geography?
- The need for fertile soil or fresh water?
- Fossil fuels, fertilizers and agricultural poisons?
- Erosion, waste and pollution?
- A negative ecological footprint?

Abundance offers a novel path for healthier people, producers and our planet.[46] The next section explores abundance methods that grow foods low on the food chain that are healthier for people, producers and our planet.

Chapter 3. What is Abundance?

Why should food producers continually extract
declining fossil resources when growers can use
plentiful resources that are surplus, cheap or free?

The design for an alternative food production system should produce food reliably, independent of weather. The process should avoid competition for finite resources with modern industrial agriculture. These resources include freshwater, fertile soil, fossil fuels, inorganic fertilizers as well as pesticides, herbicides and fungicides. Possibly the biggest challenge is the avoidance of waste and pollution.

Abundant agriculture provides a novel solution for sustainable and affordable food and energy, (SAFE) production that meets these challenges.[47] Abundance methods assure sustainability for many generations because the resources will not run out. Growers recover, recycle and reuse bioenergy and nutrients in waste streams. Of course, safeguards are necessary to assure the demise of waste stream parasites and pathogens. Growers use simple solar heaters, UV light or other technology that has been effective for over 40 years. Abundance methods cultivate algae that convert solar energy and free or low cost inputs into hydrocarbon-rich botanical biomass.

Abundance methods mimic nature's oldest and most reliable food production system at the base of the food chain. Without human cultivation, algae already synthesize roughly 80 billion tons of organic matter daily, constituting about 40% of the total fresh organic matter grown on our planet.[48] Each day, algae supply 70% of the world's oxygen, more than all the forests and fields combined. Fortunately, algae's many hungry consumers eat most the new biomass daily or the earth would be covered in algae.

One hundred times more animals eat algae than any other food because it is so nutritious. Algae's tiny cell size, only 5 μ, make the plant and ideal nutrient delivery system for plants, animals and humans. Algae and algae-infused crops offer 300 to 500% higher "nutralence" – nutrient availability and density. The tiny yet nutralent cells are immediately absorbed and bioavailable.

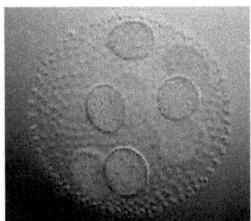

Algae

Abundance takes its name from the inputs, which are plentiful and often free, surplus or cheap – sunshine, CO_2 and wastewater. Abundance growers transform low cost inputs to cultivate valuable products. Abundance growers can clean polluted water and air while creating carbon neutral and water neutral food and energy. Every pound of algae sequesters two pounds of CO_2. The only thing released by abundance methods is pure oxygen.

Growers cultivate algae and other microorganisms in microfarms. Microfarms are flexible microcrop platforms that produce foods from plants lower on the food chain than modern industrial crops. Extensive research shows that foods low on the food chain, such as algae, provide superior nutrition for people, animals and plants than modern industrial foods higher on the food chain.

Algae, and the other microorganisms algae attract, grow protein and other nutrients 20 to 30 times more productively than land plants. This means a corn farmer would have to farm one acre for 30 years to produce as much protein as an abundance grower cultivates in one year. Growers can double the biomass daily allowing growers to harvest half the biomass every day the sun shines, year round.

Earthrise, an algae producer near the Salton Sea in southern California, has been cultivating algae for over 30 years in open raceway ponds. Earthrise produces about 500 tons annually of spirulina in 30 one-acre raceways that are about 12" deep. Spirulina is digestible and contains about 70% protein. Earthrise packages the spirulina as a healthy nutrient supplement. The company grows only for seven months a year, during maximum productivity. Some algae companies are experimenting with winter cultivars.

Cultivating Algae at Earthrise

Algae flour can substitute for corn, wheat, rice, soy or other food grain.[49] Algae oils offer a healthier alternative to vegetable or animal oils because they are lower in fat and cholesterol. Algae oils are significantly higher in micronutrients, vitamins and antioxidants than vegetable or animal oils and deliver omega-3 fatty acids.

Algae offer extraordinary biodiversity. Only about 35,000 algae species have been characterized. Experts estimate the total number of marine and terrestrial algae species to be in excess of 10 million. Each algae species produces a unique combination of protein, lipids, carbohydrates and specialty compounds. Algae culture collections available at several universities and institutes offer searchable algae lists for targeted compounds.

Algae distribution

Algae thrive all over the planet, including under the North Pole, under glacial ice in Tibet and under ice sheets in the Alps. Millions of tons of microalgae grow in both poles and form the base of the food chain for krill and other voracious algae feeders.

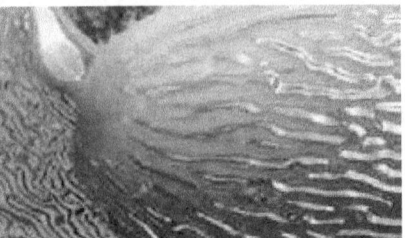

Kelp can grow a meter a day and grow to 60 meters, (180 feet)

Algae crusts form a matrix in the hottest deserts that holds soil and minimizes erosion. Algae crusts provide the structure and nutrients that support plant germination and growth in desert environments. Therefore, several algae species can be found locally or adapted to nearly any climate.

A handful of local dirt or cup of water may contain over a trillion cells and more than 100 species of indigenous algae. Indigenous algae typically out produce species from distant algae collections.[50] Local algae evolved over eons and adapted to the local microclimate.

Growers can bioprospect local algae species to find robust varieties that can be grown for protein or targeted compounds for food, feed, fabrics, fertilizers, biofuels, nutraceuticals, cosmeceuticals, vaccines, pharmaceuticals or medicines.

Weather

Abundance growers cultivate food nearly independent of climate or weather. Algae have no true growing season, although most species hit maximum productivity from spring through autumn. Cold weather species grow in the winter, although many slow or go dormant in extremely cold weather.

Covered or closed microfarms allow growers to produce algae food in practically any climate, altitude, latitude or geography. Growers in extreme locations may need to supplement with high-efficiency LED lights when sunshine is insufficient.

Open microfarms are modestly sensitive to weather. An acre raceway that typically produces 150 pounds of biomass a day may produce only 40% during a storm or on cloudy days. Unlike land plants, when conditions are not ideal, algae simply rest. As soon as sufficient sunlight appears, production returns to normal. Since growers can harvest year-round, stormy and cloudy days have only modest impact on total production.

Direct sunlight may be too strong for algae. Most algae can use only about 0.1 the amount of light the cells receive from direct sunlight. Growers often shade or diffuse sunshine to maximize production.

Open raceway and a closed system

Microfarms in a controlled environment such as a greenhouse with optional LED lights are weather independent. These growers use sunlight when available and LED lights when solar energy is insufficient. Several microfarm vertical designs capture solar energy in the top culture and use LED lights to augment sunlight on lower layers. Some growers use highly efficient LED lights to extend biomass growth into the night. Several algae companies are experimenting with 24/7 algae production with LED lights.

Cold climate growers may grow cold tolerant algae in the winter. In Canada, Sweden and other cold regions, growers add heat to accelerate biomass growth. Heat and light energy may come from renewable energy such as solar, wind, waves or geothermal.

Water

Algae thrive without freshwater. Cultivation can use non-potable water sources including brine, agricultural, municipal or industrial wastewater or ocean water. These water sources typically contain too much salinity or other pollutants for terrestrial crops. Wastewater must be cleaned with UV light, solar heaters or other tools to remove bacteria and pathogens before growing algae foods.

Land plants evolved from algae 500 million years ago. The first land plants probably used algae crusts for support as they developed their tiny root system. Roots severely limited terrestrial plants because roots made them immobile and dependent on sufficient moisture and nutrient availability within the rhizome.

Land plants die quickly with insufficient moisture because water carries the nutrients needed for cellular metabolism. If too little of any of the 24 essential nutrients are unavailable, land plants die or fail to produce seeds, so they cannot reproduce.

Algae grow to the limit of nutrients. If insufficient water or nutrients are available, algae take a rest and simply enter dormancy. When good growing conditions return, algae resume normal growth.

We have cracked open rocks thousands of years old to find algae cells on bioprospecting trips at Arizona State University. When cultured, these old cells begin their high velocity reproduction. Algae's dormancy strategy allowed these plants to sustain growth for 3.7 billion years in all types of extreme environments.

Land plants die in the presence of salt due to a plumbing problem. Salt ions in solution are too large for root absorption. The clog in the roots blocks water flow and causes the plant to starve. Algae evolved in ancient oceans, which were extremely saline. Algae have no roots and absorb nutrients directly into the cell, independent of the in-situ saline concentration.

Brine water makes up half the groundwater stored on the planet. Therefore, brine water alone provides a sustainable cultivation medium for algae foods. Brine water typically has lower salinity than ocean water. When crude oil pumps remove oil from the ground, they often recover 20 barrels of brine water for every barrel of oil. Today, this nutrient rich water is pumped deep into the ground to create pressure this is the removal of additional crude oil. Soon, algapreneurs will cultivate algae that remove the nutrients before the brine water is returned to the well. In many areas, brine water lies a few feet under the ground and can be recovered by foot pumps.

In most cities, wastewater creates a huge cost. Algae can transform those costs to positive revenue, creating valuable biomass while cleaning the wastewater. The oldest algae application in the U.S. is wastewater treatment.

Many freshwater algae species can be trained to grow effectively in brine or wastewater. Many sea vegetables, edible seaweeds, also thrive in brine and saline wastewater.

Sea Vegetable – Edible Seaweeds

Waste

Industrial agriculture produces massive waste and pollution. Abundance methods recover and repurpose bioenergy and nutrient waste into energetic foods with high nutralence. Abundance growers can clean polluted air, water and soil while they produce valuable algae biomass.

Currently, waste streams represent a huge cost for farmers and zoos. Zoos often have to pay more to dispose of the ZooPoo than they pay for animal feed. ZooPoo includes animal, botanical and trash wastes. The zoo experiences huge costs for handling, storing, inspecting, hauling and disposing of the ZooPoo. Currently, ZooPoo adds tons of biomass to landfills, which degrade the environment.

The amazing thing about ZooPoo is its value. Animal manure and botanical wastes retain roughly 60% of the bioenergy that was originally in the plant. Even more importantly in an economic sense, ZooPoo retains 80 to 90% of the original plant nutrients.

Abundance methods can recover and reuse the zoo waste stream and transform this huge cost to a profit center. The combined value of the reclaimed bioenergy and nutrients could cover the cost of animal care. A smart zoo will create an additional ZooPoo revenue stream with a world-class ecotourism exhibit that conveys how algae support sustainable green living.

Human, industrial and agricultural waste streams often contain heavy loads of fertilizers, pesticides, herbicides and fungicides that make them unacceptable for use on food crops. Tiny algae cells absorb individual elements from complex compounds such as pesticides, which detoxifies the agricultural poison. Current technology offers several algae solutions that detoxify waste streams.

Drugs create a serious problem with manure as fertilizer. Organic food producers want to use animal manure as fertilizer but in many cases they cannot due to the pharmaceutical drug problem. For 60 years, meat producers have fed antibiotics to farm animals to increase their growth and prevent infections. Nearly 70% of all antibiotics produced in the U.S., nearly 25 million pounds a year are fed to cattle, pigs and poultry according to the Union of Concerned Scientists.

Feeding pharmaceuticals to animals sustains a growing demand for meat but it generates public health fears associated with the expanding presence of antibiotics in the food chain. Over 90% of the drugs in animals and humans end up being excreted either as urine or manure. Food crops absorb and concentrate antibiotics and other drugs when grown in soil fertilized with livestock manure.

Municipal and industrial waste streams often contain additional pathogens, pharmaceuticals and poisons. Algae offer solutions to detoxify these waste streams and to recover the nutrients.

Abundant agriculture provides air pollution solutions. Power, cement and manufacturing plants produce the heaviest load of carbon dioxide to the atmosphere. Several algae producers such as Carbon Capture Corporation are designing systems that sequester carbon from exhaust plumes. Current technology allows microfarms to capture only part of the waste stream because power plants operate 24/7 while algae grow during daylight hours. Since every ton of algae captures two tons of CO_2, waste plumes provide substantial economic benefit. Smaller microforms may locate near other carbon sources such as restaurants, cleaners or breweries.

Companies like Biovantage Resources are innovating with water pollution solutions. The use of algae to clean wastewater is the oldest algae application in the U.S. New algae technologies are making water reclamation more effective and efficient. Several companies have business models where their primary revenue stream comes not from the algae produced but recovery and resale of metals and nutrients. Biovantage Resources is currently installing and wastewater treatment system for a new gas recovery operation in North Dakota. The system uses solar energy augmented by LED lights to clean the water and produce algae products.

A dairy farmer in California must pay $0.35 per cow each day to dispose of liquid and solid wastes. The waste disposal cost for a 10,000 cow farm penalizes the farmer $35,000 a day. Farmers typically use the cheapest disposal method, which is burning or burying the waste.

Research at St. Cloud University in Minnesota has used algae to recover the residual energy from cow manure. This research suggests that the bioenergy value from each cow's manure is higher than the value of the milk. This calculation does not include the substantial savings from manure disposal.

Abundance products

Abundance methods can produce practically anything sourced today from land plants because land plants evolved from algae 500 million years ago. Most fossil fuels are made of algae or algae feeders fossilized from ancient oceans. Therefore, anything that can be made from fossil fuels can be made from algae. Algae biofuels burn without black smoke particulates because they missed the hundreds of millions of years in fossilization. Algae biodiesel burns cleanly, similar to vegetable oil.

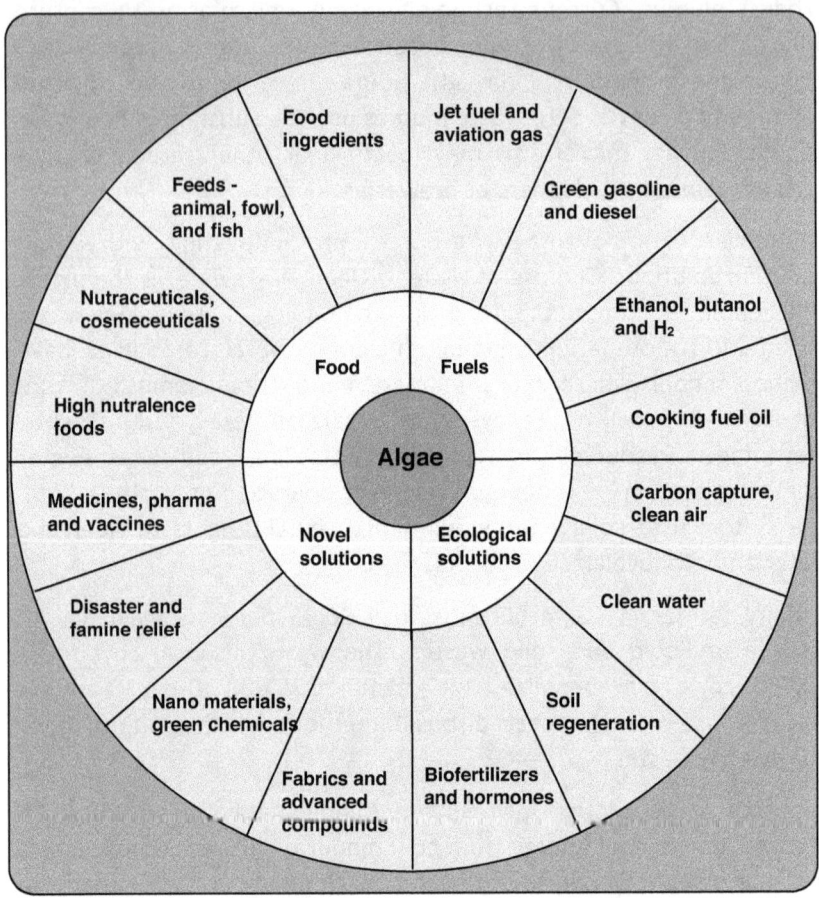

Figure 3.1. *Algae Products and Ecological Solutions*

Other sources expand on the many entrepreneurial opportunities with abundance production in each sector.[51]

Chapter 4. What are Freedom Foods?

Imagine superior fossil-free foods that clean rather than pollute our ecosystems.

Freedom foods grown with abundance methods offer the opportunity to provide a food source that augments but does not compete for declining fossil resources with modern industrial foods. [52] Abundance produces healthier food independent of weather and freshwater while cleaning air, water and soils. Growers can produce these foods 20 to 30 times more productively than industrial foods.

Freedom foods reinvent our food supply from the foundation of the food chain and:

- Free consumers for smart choices for healthier and tastier food.
- Free growers from consumption of fossil resources.
- Free ecosystems of waste and pollution.
- Leave the environmental footprint of a butterfly.

Foods grown low on the food web require a fraction of the energy consumed by modern food. They are free of the resource consumption and waste caused by industrial crops because they are grown organically, with no or minimal fossil resources.

Note: Freedom foods include the full spectrum of microcrops such as algae, yeast, fungi, bacteria, archaea, protists, plankton and many others that grow in symbiotic communities.

This alternative food supply provides splendid natural produce and does not compete with industrial agriculture because growers use predominately bountiful resources. Growers recycle waste stream nutrients that are plentiful, affordable and will not run out, Figure 4.1.

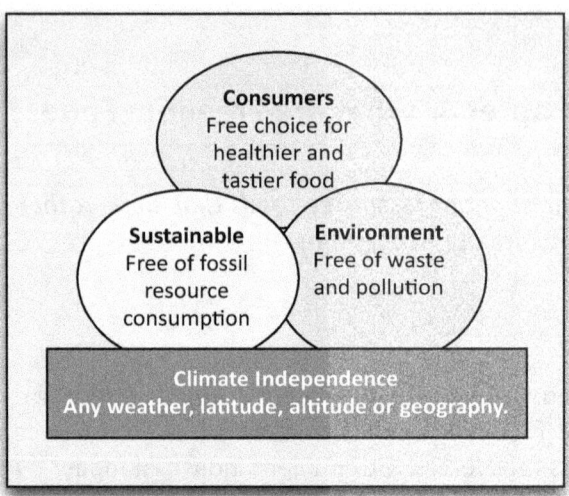

Figure 4.1. *Freedom Foods*

Scientific evidence and common sense show that foods low on the food chain consume significantly fewer resources. Consider the fresh water consumption of one pound of:

- Beef   26,400 gallons
- Corn   168 gallons
- Algae   0 gallons

A pound of algae provides twice the protein than beef and three times the protein of corn. Algae provide 50 to 500% more nutralence, essential nutrient availability and density. Micronutrients include vitamins, minerals, antioxidants and trace elements. Since algae cells are so tiny, they are immediately bioavailable to the body, which aids both digestion and nutrient assimilation. Bioavailability increases nutrient absorbtion, allowing consumers to eat less while gaining more essential nutrients.

What are Freedom Foods?

Freedom foods are clean, healthy, and extremely low in fat and cholesterol. Consider the value proposition for a chocolate cake.

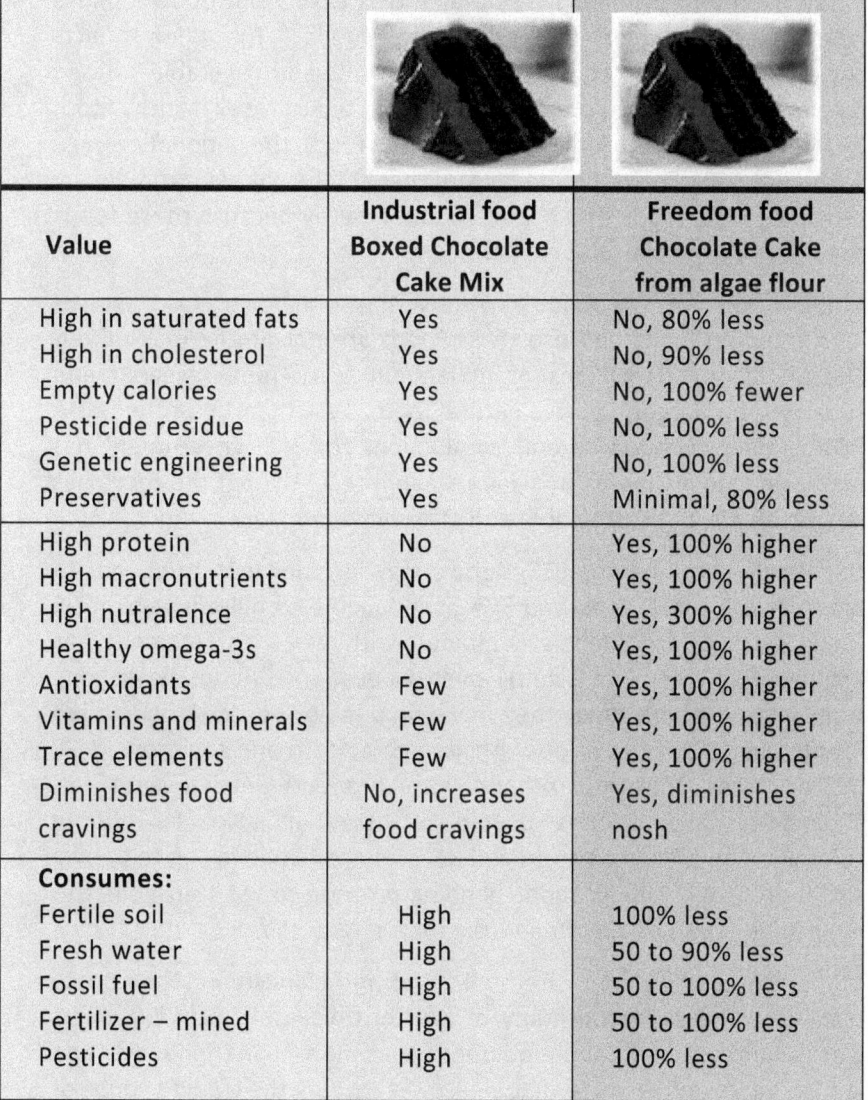

Value	Industrial food Boxed Chocolate Cake Mix	Freedom food Chocolate Cake from algae flour
High in saturated fats	Yes	No, 80% less
High in cholesterol	Yes	No, 90% less
Empty calories	Yes	No, 100% fewer
Pesticide residue	Yes	No, 100% less
Genetic engineering	Yes	No, 100% less
Preservatives	Yes	Minimal, 80% less
High protein	No	Yes, 100% higher
High macronutrients	No	Yes, 100% higher
High nutralence	No	Yes, 300% higher
Healthy omega-3s	No	Yes, 100% higher
Antioxidants	Few	Yes, 100% higher
Vitamins and minerals	Few	Yes, 100% higher
Trace elements	Few	Yes, 100% higher
Diminishes food cravings	No, increases food cravings	Yes, diminishes nosh
Consumes:		
Fertile soil	High	100% less
Fresh water	High	50 to 90% less
Fossil fuel	High	50 to 100% less
Fertilizer – mined	High	50 to 100% less
Pesticides	High	100% less

Table 4.1. *Industrial Food versus Freedom Food Chocolate Cake*

Freedom to choose healthy food

Many modern consumers do not have access to healthy foods. The Centers for Disease Control and Prevention published the Modified Retail Food Environment Index that shows that 9 out of 10 families lack access to retailers that sell healthy foods.[53] The Index reflects consumer access to retailers with fresh fruits and vegetables. Based on a range from zero (no food retailers that typically sell healthy food) to 100 (only food retailers that sell healthy food), the national average score was 10. Freedom foods can change the access problem by providing local foods that are fresh and healthy because these foods can be produced almost anywhere.

Modern consumers cannot currently make healthier food choices with freedom foods because these foods are not on the market yet. The USDA spends billions of dollars on fossil food research and subsidies, but largely ignores natural foods. Less than 1% of the USDA R&D budget supports organic production. The U.S. government has made small investments in algae research recently, but the focus has been biofuels, not sustainable and affordable food.

Genetically engineered, (GE) foods may or may not bring health hazards. The USDA, FDA, and EPA have failed to require labels on GE foods. Lack of food labels combined with weak enforcement has enabled "GE creep." In slightly over a decade, GE foods have gone from zero to where today they make up a major portion of packaged foods. Over 90% of U.S. food grains such as corn and soy grow in GE monocultures that are refined into products loaded with fat, cholesterol, and calories devoid of essential nutrients, (empty calories). The USDA has approved 81 GE crops, while failing to deny a single proposal.[54] Applications pending propose to use transgenics to alter up to 30 genes simultaneously for a single crop.

Modern processed foods are high in fat and cholesterol. They cause obesity, diabetes, and a litany of Western diseases, including heart disease and cancers. Childhood obesity has more than tripled over the last 30 years and leads to diabetes.[55] Diabetes is the leading cause of kidney failure, lower-limb amputations, and new cases of blindness in the U.S. Diabetes is a major cause of heart disease and stroke and the seventh leading cause of death in the U.S.[56] The American Diabetes Association estimates the cost of diabetes in the U.S. exceeds $174 billion a year.[57]

Freedom foods are healthy and can treat and in some cases prevent obesity, diabetes, and other diseases. Not only are they low in fat and cholesterol but they stop empty calories with high nutralence. These foods are naturally biodiverse, which eliminates the need for GE monocultures. Unlike fossil foods, freedom foods are clean, free of chemical fertilizer and pesticide, herbicide, and fungicide residue.

Organic foods

Organic foods may be healthier for people and producers than industrial, but represents less than 3% U.S. of cropland.[58] Organic farmers avoid, to the degree possible, GE seeds, chemical fertilizers and agricultural chemicals and poisons. Unfortunately, organic farming methods cannot meet global food needs because organic production uses more fossil resources than industrial agriculture. Organic production creates substantially more food waste due to increased penetration of field pests and shorter produce shelf life.

Organic production offers some resource advantages compared with industrial farming but gives no more weather tolerance than other forms of agriculture. Organic farmers often must use more cropland than industrial to produce the same amount of produce. Organics are equally consumptive of water and often higher in fuel consumption. Organic farmers do an excellent job at sustaining natural soil fertility but at the high time and fuel cost of collecting, transporting, storing, applying and turning compost into the soil. Organic producers cannot use no-till methods because the organic compost or fertilizer must be plowed into the soil. Compost left on the soil surface loses much of its nitrogen as it turns to gas in a process called volatilization,

Freedom foods can use organic or industrial production methods. Ideally, growers use abundance methods that are organic and, in addition, use no or minimal fossil resources. Abundance growers may use waste, ocean or brine water, which is sustainable for many generations.

Natural path

Freedom foods follow nature's path, exploiting the oldest and most efficient food production system on Earth. Algae and the diverse microorganism communities algae attract provide superior nutrition for people, animals, and plants. The tiny algae cells are packed with nutrients and are immediately bioavailable to consumers.

Corn produces its first gram of protein in about 120 days, a full growing season. Consumers and growers must wait another 365 days for the next harvest. Freedom foods growers produce the first gram of protein in about two weeks. Growers then harvest additional protein every few days, all year round.

Microflora communities are similar to the organisms in our gut that break down food and aid digestion. Growers cultivate microcommunities in microfarms and train them to produce food, feed, fertilizer, and many other coproducts. The focus here is on algae, but a diversity of microflora may yield similar good foods with attributes superior to fossil foods.

These tiny biofactories run efficiently as they recycle and reuse the residual energy and nutrients from brine water or waste streams. Microfarmers use solar heaters and UV light to kill pests and pathogens before algae go to work to absorb wastewater organics.[59]

Higher nutralence

Succulence is the natural ability of succulent plants to absorb and hold water.[60] Algae demonstrate nutralence as the biomass concentrates nutrients at substantially higher levels than land plants.[61] The reach of their roots and the available soil nutrients limits the nutrient density of land plants. Algae avoid the root problem by living without roots.

Algae provide a low fat, low calorie, nearly cholesterol-free source of protein. Some algae, such as spirulina, contain up to 70% protein by dry weight − twice the protein of meat. Unlike meat, most algae varieties provide the full complement of nine essential amino acids. The low fat content, only 5-10%, is a fraction of other protein sources.

What are Freedom Foods?

One tablespoon, 10 grams of algae delivers the same amount of:

- Calcium as 8 tbs milk, 32 tbs soybeans, 8 carrots, or 22 tomatoes.
- Magnesium as 40 tbs milk, 9 carrots, or 6 tomatoes.
- Iron as 512 tbs milk, 8 tbs soybeans, 11 carrots, or 5 tomatoes.

Field studies show that algae nutralence, other vitamins and minerals are similarly 200 to 300% denser than field crop produce.[62]

Algae offer more calcium per tablespoon than 22 tomatoes

The fibrous components of algae add bulk to the digestive tract reducing hunger pangs, transit time, and intestinal pathologies.[63] The total fiber content of algae (~6 g/100g) is greater than that of fruits and vegetables promoted today for fiber content: prunes (2.4 g), cabbage (2.9 g), apples (2.0 g), and brown rice (3.8 g).[64]

A chicken egg contains about 300 mg of cholesterol and 80 calories while providing the same protein as a tablespoon of the algae spirulina, which carries 1.3 mg of cholesterol and 36 calories. Algae are also an excellent plant source of glutamic acid, an amino acid that promotes intestinal health and immune function.

Each kilogram of algae biomass has two or three times the protein available from a kilogram of food grain. Algae concentrate many other nutrients at a multiple of the nutrients found in grains. Foods made from algae offer substantially more nutralence than food grains. Consumers benefit with more nutrients per calorie and per bite.

Per tablespoon, algae provide 10 times the beta-carotene available from the land plant that delivers the most beta-carotene, carrots. Beta-carotene are carotenoids that are highly pigmented (red, orange, yellow), compounds naturally present carrots and some other vegetables and fruits. Alpha, beta, and gamma carotene are considered provitamins because they can be converted to active vitamin A. Carotenes possess antioxidant properties and serve vital biological functions. Vitamin A deficiency leads to abnormal bone development, disorders of the reproductive system, xerophthalmia (dry eyes), night blindness and ultimately death.

Algae hold 10 times more vitamin B-12 and iron than beef liver. Vitamin B12 is an essential vitamin found in some fish, dairy products and animal liver. B-12 deficiency causes pernicious anemia, the inability to absorb vitamin B12 from the intestinal tract. B12 deficiency is common in the elderly, HIV-infected persons and vegetarians who are not getting sufficient B12 from their diet.

Natural biodiversity

Abundance growers have access to splendid natural diversity that enables them to grow biomass of 30 to 70% protein, depending on their target food or coproduct. Growers that want to maximize lipids (oils) may select a species that contains 40% lipids. Other growers may want to maximize production of carbohydrates, pigments, vitamins, minerals, antioxidants, cosmetics, medicines, vaccines or many other valuable coproducts.

Algae absorb a wealth of mineral elements that concentrate as about one third of its dry biomass. The macronutrients include sodium, calcium, magnesium, potassium, chlorine, sulfur and phosphorus while the micronutrients include iodine, iron, zinc, copper, selenium, molybdenum, fluoride, manganese, boron, nickel and cobalt.

Health benefits

Although very low in fat, algae offer an excellent source of the essential polyunsaturated fatty acids. The omega-6 and omega-3 fatty acids (ARA and EPA/DHA respectively) are necessary for normal metabolism, as they are the precursors to critical hormone-like, signaling molecules known as the eicosanoids. These short-lived messengers direct life-supporting functions such as blood clotting, inflammation, vasodilation, blood pressure and immune function. Only small amounts of ARA and EPA/DHA are needed daily (<1 g), and one tablespoon of algae can supply about half this amount.

Fish do not synthesize omega-3 in their oil. Algae synthesize the omega-3s that fish accumulate in their oil and that support human brain, eye and heart functions. Algae-based foods provide vital polyunsaturated fatty acids, omega-3, 6 and 7.

Medical research shows that omega-3 fatty acids reduce inflammation and may help lower risk of chronic diseases such as heart disease, cancer, and arthritis. Omega-3 fatty acids concentrate in the brain and are important for cognitive (brain memory and performance) and behavioral functions. Infants who do not get enough omega-3 fatty acids from their mothers during pregnancy are at risk for developing brain, vision and nerve problems. Insufficient omega-3 deficiency include fatigue, joint pain, poor memory, dry skin, heart problems, mood swings and poor circulation.

Algae contain a wide spectrum of prophylactic and therapeutic factors that include vitamins, minerals, amino acids and essential fatty acids. Algae provide the super anti-oxidants such as β-carotene, vitamins A, B, B-complex, C, D, E, and K, and a number of unexplored bioactive compounds.[65] These constituents stimulate numerous metabolic pathways and promote antioxidant, anti-bacterial, antiviral, anticancer, anti-inflammatory, anti-allergic, and anti-diabetic actions. Extensive medical research shows algae constituents promote vascular, mental, and intestinal health.[66]

Phytic acid compromises the mineral availability from land plants, particularly legumes and grains, because the acid binds the minerals, rendering them unavailable for absorption into the blood stream. Phytic acid is typically absent in many algae species. Studies show that iron absorption is 3.5 fold greater for algae compared to rice.[67] Algae iron is easily absorbed by the human body because its blue pigment, phycocyanin, forms soluble complexes with iron and other minerals during digestion making iron more bioavailable. Hence, unlike iron derived from land plants, the bioavailability of algae iron is comparable to that of heme iron in meats.[68]

Algae nutrients, vitamins and minerals enhance physiological systems including the cardiovascular, respiratory and the nervous systems.[69] Algae components activate the cellular immune system including T-cells, macrophages, B-cells and anti-cancer natural killer cells.[70] Algae polysaccharides inhibit replication of several enveloped deadly viruses including herpes simplex, influenza, measles, mumps, human cytomegalovirus, SARS, and HIV-1.[71]

Algae's nutralence, antioxidants, enzymes and extracts, boost the immune system and enhance the body's ability to grow new blood cells. Algae are rich in phytonutrients and functional nutrients that activate digestive and immune systems. Algae compounds accelerate production of the humoral system (antibodies and cytokines), enabling the body to protect against invading germs.[72] Specific algae polysaccharides have demonstrated anti-atherosclerotic functions, reducing blood LDL cholesterol concentrations, and cardiovascular disease risk.

Research on humans and animals shows algae components have utility in the prevention and control of diabetes.[73] Other studies have demonstrated algae's therapeutic value for cholesterol management, blood pressure, heart disease and cancers.[74] Algae can moderate chronic inflammation that often precedes degenerative diseases. Algae provide therapeutic value for diabetes and fat metabolism.[75]

Research on mice shows algae delays the onset of motor symptoms and disease progression in ALS (Lou Gehrig's disease). Algae reduce inflammatory markers and motor neuron death.[76] Algae are calcium rich and may protect against osteoporosis.[77] Recent research suggests algae activate human stem cells, which provide a spectrum of health benefits, including moderation of brain degeneration.[78]

New food supply

Freedom foods can transform our food production system so that consumers could make healthy choices for themselves, producers, our planet and our atmosphere, Figure 4.2. These foods can be grown locally or regionally and are superior in nutrition and taste, yet create minimal pollution or waste. This new food supply will enable us to leave a positive legacy for our children – healthy, affordable food, clean ecosystems, breathable air, and abundant natural resources. Freedom foods offer a clean, naturally biodiverse and healthy alternative to fossil foods.

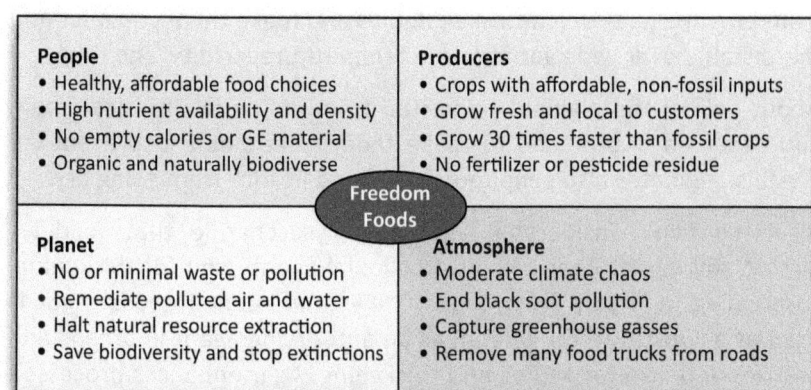

People
- Healthy, affordable food choices
- High nutrient availability and density
- No empty calories or GE material
- Organic and naturally biodiverse

Producers
- Crops with affordable, non-fossil inputs
- Grow fresh and local to customers
- Grow 30 times faster than fossil crops
- No fertilizer or pesticide residue

Freedom Foods

Planet
- No or minimal waste or pollution
- Remediate polluted air and water
- Halt natural resource extraction
- Save biodiversity and stop extinctions

Atmosphere
- Moderate climate chaos
- End black soot pollution
- Capture greenhouse gasses
- Remove many food trucks from roads

Figure 4.2. *Freedom Foods benefit everyone*

The use of plentiful resources that will not run out presents the opportunity for food democracy, where everyone has access to good food or the affordable inputs to grow their own food. The use of plentiful resources means food supplies will be available and affordable for many generations.

Freedom Foods vision:

> *Distribute the knowledge and capability for abundance methods globally to enable all people to grow good food and coproducts for their family and community locally.*

Distributed, local food production produces fresh foods that do not need preservatives and avoid the high dollar and energy cost of transportation.

Algae in the food supply

Most consumers are surprised that food processors already integrate algae components in our food supply. A market basket test at Arizona State University examined the non-fresh produce items in 10 typical shopping carts for foods containing algae components. About 72% of the foods and 88% of the cosmetics contained algae constituents.

Beer and soft drinks use algae as a clarifier. Algae cell walls contain carrageenan used as a stabilizer or emulsifier found in dairy, confectionary and bakery products. Alginates provide alginic acid from brown algae, which thicken liquids and make them creamier and more stable over wide variations in temperature, acidity, and time.

About half of the alginate harvested from kelp goes into ice cream and other dairy products to make them smoother and prevent ice crystals. Alginates also keep toothpaste and lipstick from going dry.

Algae contain considerable agar, a polysaccharide that solidifies almost any liquid. Agar acts as a colloidal agent used for thickening, suspending, and stabilizing soups, stews and canned goods. Agar is used as a substitute for gelatin, as an anti-drying agent in breads and pastries and also for gelling and thickening. Agar enhances processed cheese, mayonnaise, puddings, creams, jellies, and dairy products.

Algae components make delicious Freedom Foods

Algae naturally form flours that can substitute for food grains such as rice, corn, wheat or barley. Algae flour makes chips, dips, breads, tortillas, crepes, cakes and pretzels that have superior nutrient profiles to their cousin foods with substantially lower fat and cholesterol. Of course, people will not choose the algae-based food models unless they taste similar or better and are affordable.

Algae milk can substitute for dairy, soy, or almond milk while providing higher protein. Many people are lactose intolerant and others are bothered by soy and nut allergies. Algae milk induces neither intolerance nor food allergies. Algae sugars can substitute for cane, beet or corn sugars. Solazyme recently announced new algae chocolate improved mouth feel, yet 85% fewer calories.

Texturized algae made into vegie burgers pack the savory umami taste. Currently, the extremely popular Umami hamburger chain in

What are Freedom Foods?

Los Angeles must add MSG to their beef to gain the attractive hearty flavor. When the Umami firm uses algae, they will be able to provide great flavor with natural savory taste from algae.

Consumers in a few years will have a choice for hamburger. A beef hamburger, with supply costs and environmental takes, may cost $35. A cultured meat burger, where the meat is cultured in a laboratory without and animal, may cost $25. (Currently, cultured meat has the consistency of snot.) The algae burger alternative probably will cost about $8 and offer all the health benefits associated with other freedom foods. Consumption of the algae burger will allow the eater to leave a net zero ecological footprint, similar to a butterfly.

Availability

Sea vegetables and other algae products are popular for their nutrition, color, taste and texture in Asian Pacific Rim countries.

Sea Vegetables

Many high-end European restaurants offer algae appetizers, entrees, main courses and deserts. In the U.S., freedom foods are available currently only in Asian markets and health food stores, where they sell as food supplements. Whole Foods, Trader Vic's and Costco recently began selling algae snack foods.

Roasted Seaweed at Costco

Food companies have investigated algae-based foods for decades but they faced a pair of showstoppers with freedom foods – supply and demand. Growers have no incentive to grow algae when terrestrial foods are so heavily subsidized and cheap. Consequently, food-processing companies have no suppliers.

General Foods, Nabisco, Frito-Lay, Borden, Dial or Quaker Oats cannot manufacture an algae product today because there is no market. Consumers do not select these foods because they know little about the health and environmental value of freedom foods.

Algae in restaurants

Why does the number one rated restaurant in the world, NOMA, serve algae? Rene Redzepi chef, forager, and owner of NOMA, in Copenhagen, made the March international cover of *Time Magazine* for his excellent innovations in locally sourced and foraged cuisines. His extraordinary NOMA cookbook shows beautiful pictures of algae garnishing and supporting his world-class servings.

Rene Redzepi and Umami Burger

The human tongue has five taste buds: sweet, sour, bitter, salty and umami. Umami, the savory or hearty taste comes from taste buds in the middle of the tongue that give a savory or hearty taste – not to just the food eaten directly, but also to the accompanying food. Many upscale restaurants offer algae directly or accompanying hors d'oeuvres and main courses to add the desirable savory taste. One of the most popular new restaurant chains in Los Angeles, Umami Burger, has found great success promoting the umami taste.

What are Freedom Foods?

Dozens of excellent algae recipes and algae foods are available at www.AlgaeCompetition.com and on Epicurious.com

Algae flour

Solazyme in partnership with Roquette Nutritionals announced in April 2012 the availability of their Algalin™ flour. Algalin flour looks and acts like flour, but is actually a lipid substitute that is similar to olive oil. Product recommendations explain that the flour can substitute for eggs, cream, milk, vegetable oils or other lipid sources.

Algalin flour provides improves nutritional profiles in many foods such as bakery, beverages and frozen desserts. Algalin flour, made from chlorella, acts as a whole food ingredient and delivers very low saturated fat, no trans-fats and no cholesterol. It reduces calories considerably, as well as provides fiber and protein, while providing the same overall mouthfeel and consistency as a full fat food.

Algalin™ brownies, flour, oil and nutritional drinks

Altein™ Algalin protein provides over twice the levels per gram as corn or soy protein. The product has about 20% dietary fibers, 10% lipids and a wide array of trace minerals and micronutrients. The vegan product is marketed as gluten-free, non-allergenic, sustainably produced with highly digestible protein.

Since the Solazyme algae products are made with fermentation and require sugar for the energy instead of photosynthesis, some may argue the sustainability claim. However, Solazyme works very hard on their sustainability and has invested in farms to grow plants that can supply the necessary sugars.

New demand for substitutes

One of the most interesting recent food stories belongs to Girl Scouts and palm oil. Two Girl Scouts created an initiative to end the use of palm oil in Girl Scout cookies because palm oil farming causes rainforest deforestation endangers thousands of animal species and contributes to human rights abuses. Over 70,000 people have signed their petition to stop the use of palm oil in Girl Scout cookies. The two girls have been featured on numerous news and talk shows, and they were recently honored with the United Nations Forest Heroes Award for their work in saving rainforests.

Two Girl Scouts work to end Palm Oil use in Scout Cookies

Fish oil may be right behind palm oil. Several environmental and scientific groups are working to end the unsustainable catch of small fish for animal feed and fish oil, omega-3 fatty acids. Two independent studies recommend cutting the small fish catch by 50%. Some fisheries are already fished out. The island of Sardinia has no more sardines because overfishing caused the fishery to crash.

"Little Fish, Big Impact," financed by the Lenfest Foundation through the Pew Charitable Trusts, reports how small forage fish catch has increased to 31 million metrics tons or 37% by weight of all fish harvested globally.[79] Thirteen eminent scientists invested three years reviewing the scientific literature and analyzing computer models of food webs in 72 marine ecosystems around the world. They found that 75% of the ecosystems they studied had one or more large predators for which forage fish made up at least half their diet.

Forage fish are algae feeders that deliver essential nutrients absorbed and synthesized by algae to their consumers higher in the food web.

Forage fish are a critical link in the food chain and are consumed by just practically every fish, bird and mammal in the ocean, including tuna, dolphins, puffins and penguins.

The task force at the Institute for Ocean Conservation Science at Stony Brook University led by Ellen Pikitch estimated that forage fish are worth more than $11 billion to the oceans. Left in the ocean, forage fish are worth twice as much as when processed for fish oil and aquaculture.[80]

As companies look for sustainably sourced oils, algae oils will become popular rapidly because the value proposition has already been communicated to consumers: "save dolphins and puffins."

Save dolphins and puffins, eat algae omega-3 oil

Commodity prices

Constantly rising commodity prices are prompting demand for cheaper and more sustainable substitutes for terrestrial crop commodities. Commodities that have been relatively constant for years are rising.

Higher commodity prices translate quickly into higher food prices. The Arab Spring ignited over high food prices and limited food availability in Tunisia, Libya, Yemen and Egypt. Food prices in the region exceed 40% of family income. Global food demand is high and rising fast. Reserves are so low that price sensitivity to crop yield losses has become extreme. As food prices continue to rise, more countries will face uprisings that escalate to revolution.

Algae can play a major role in sustainable cultivation of substitutes for food grains, vegetable oils, fish oils, and other sources of protein, including meats. Business opportunities abound in the algae realm of sustainable food and energy products.

Freedom foods adoption

Skeptics predict that consumers will not accept freedom foods, despite their many advantages. They argue that consumers do not like change; especially in the foods they eat. The science of consumer behavior offers two strategies to change people's food consumption:

1. **Push** people to eat lower on the food chain with influence strategies, such as carefully explaining the health and environmental benefits.

2. **Pull** occurs as consumers demand and buy nutritious and delicious foods that appear similar, yet taste better than their equivalent fossil food.

The push strategy called cognitive conditioning, works only for a tiny percentage of consumers. Most people try diets or practice good food habits for a brief period, and then become recidivists. They revert to their favorite comfort foods. The statistics on diets confirm that most do not work.[81] Food behaviors are central to family, culture and society, and are extremely difficult to change.

Push strategies have failed scientists and practitioners repeatedly. Many obese people are experts on nutrition. Repeated studies show that knowledge alone fails to change behavior. People change their behavior when they find a preferable substitute.

Push strategies cannot work because fossil foods monopolize our fields, stores, refrigerators and plates today. The only choice most consumers can make currently is industrial versus organic foods or meat versus vegetables. Industrial foods control 97% of the U.S. market and 94% in Europe. Unlike the U.S., the European Union offers organic growers subsidies for the social, environmental and health benefits provided by organic foods. Freedom foods offer a fossil free alternative, aligned with organic methods, and will give consumers another choice – once they are available.

What are Freedom Foods?

Freedom foods offer a pull strategy where consumers are attracted superior foods. These foods from low on the food chain go into products that appear similar to traditional foods but with ingredients that are natural, healthier and provide superior sensory pleasures. To gain widespread consumer adoption, freedom foods must offer more than health and environmental benefits. Consumers must perceive them as offering a bundle of positive attributes. Each additional positive characteristic will accelerate consumer adoption. When consumers have the freedom to choose foods with superior sensory appeal and better nutrition at a reasonable cost, they are likely to pull freedom foods from grocery shelves.

Freedom foods will attract some consumers for the health benefits associated with eating low on the food chain. Others will choose freedom foods to avoid GE crops, preservatives and chemical and poison residues. Some will want freedom foods to lighten their ecological footprint. The majority will probably make their choice based on superior nutrition and taste.

Freedom foods will spark demand by educating consumers, which will motivate new suppliers. The new freedom foods industry will offer thousands of engaging entrepreneurial opportunities for growers, suppliers, restaurateurs and chefs.

Chapter 5. Algae in Human Food History

Did algae make us human?

Algae's appealing taste may have helped us become human by attracting our ancestors to algae and the Omega-3s that sparked brain enlargement.[82]

Algae played pivotal roles in human evolution and survival. Early human societies evolved along coastlines, rivers and lakes and depended on algae for food and medicines.[83] The high nutralence biomass was plentiful year-round and easy to harvest. Many groups ate algae directly and probably ingested algae in their drinking water.[84]

Algae provided a rich and nearly complete source of nutrition – a complex blend of nutrients that no other food source, plant or animal, could offer.[85] Algae were analogous to a modern-day vitamin supplement – but actually, algae are a more robust, natural, and inclusive blend of healthful nutrition.[86]

Algae offer superior protein, particularly the red, green and blue-green algae. Spirulina, blue-green algae contains 70% protein (dry weight), which is higher than and corn, 23%.[87] Algae protein content is highest in the late winter and early spring, which is advantageous when terrestrial plant food sources are scarce. Algae nutralence benefited our ancestors year round by preventing many of the nutrient deficiencies that plague modern human societies.

Peace Microfarms

Did Algae make us human?

Possibly the most interesting unanswered question in science is: "How did we become human while our contemporary Homo cousins became extinct?"

Our ancestor's attraction for the sweet taste of algae may have played a significant role in becoming human. Strategically placed evidence on our tongue provides fascinating clues that science has so far missed. Our pre-human ancestors made a significant, possibly accidental, decision to ingest algae, which may have led to the evolution our large brain and enabled *Homo sapiens* to evolve, thrive and rise to the top of the food chain.

Scientists agree that human brain enlargement—encephalation—differentiates *Homo sapiens* from our ancestors. Our pre-human ancestors evolved from chimpanzees around 8 million years ago (mya) but very little happened to the brain for the first 6 million years. About 2 mya, brain enlargement began and by 1.5 mya, the humanoid brain was three times the size of chimpanzees. What happened to our ancestors during this half a million years of evolution? Humanoids brains grew larger a million years before cooking fires or hunting weapons were invented.

Larger brains require substantially more energy because brain mass consumes 16 times more energy than muscle mass. Therefore, our ancestors traded muscle for brains. Something triggered brain enlargement and the logical answer was a change in diet. Current theory posits that our ancestors moved from the primate diet of leaves, bark, insects and occasionally fruit to a more diverse diet, higher on the food chain that included game meat.

However, moving up the food chain to hunt game meat would have been problematic for slow, scrawny and pitifully weak hominids that the fossil record shows had Lucy's stature of only 3.5 feet. Relative to predators, they had poor senses; including sight, hearing and smell. Had they decided to hunt, they would have quickly entered the food chain rather than dominating other animals.

Other scientists suggest early hominids practiced scavenging to gather game meat. They may possibly have found bones that could be pounded for the marrow. Since both predators and scavengers 2 mya were twice the size they are today, hominid scavengers would have been very lucky to find a carcass. The hominid scavenger probably would have run out of luck hauling the food back to camp because stealth predators were numerous and ferocious. Early hominids probably were subject to annual predation at the same rates at which living primates living under natural conditions are today—roughly 8% of their local population.

Independent of the physical improbability, the scavenging scenario is unlikely due to our weak stomachs. Meat begins putrefying immediately after death. Spoiled meat attracts parasites, insects, maggots, worms, bacteria and other microorganisms that would have been just as fatal for early hominids as they are for humans today.

Anthropological evidence shows humans evolved in East Africa along the Rift Valley Lakes that are unique ecosystems. These ancient, high alkaline soda lakes have large natural stands of spirulina, which flourishes as a nearly pure culture in the high pH alkaline water. The blue-green algae do not fix nitrogen and the biomass is safe to eat.

Moving down the food chain

The dance theory suggests that we became human with a three-step waltz. Our ancestors probably took two small steps down the food chain before taking a big step up to hunt game.

Early hominids' first step was probably down the food chain where they ingested algae in their drinking water. Terrestrial plant foods available to early Homo were largely hard, dry and bitter. Wild seeds, nuts, roots and leaves would have provided little variation in terms of the four classic tastes; sweet, bitter, salty or sour. Sweet would have been largely missing from their diet, except for the rare occurrence of fruit. Salty would have been missing except for salt licks and they provide an unpleasant intense salty taste, independent of food. The sweet, green algae water would have provided a subtle sweetness and occasionally saltiness. Compared to the rest of the diet available to early Homo, sweet green water would have been very attractive.

Of course, early hominids did not intentionally ingest algae to enlarge their brains. Evolution is not volitional and our ancestors probably could not even see the algae in the water because the cells were too small. Algae first become visible as a light cloud. As the cells proliferate, they turn the water green but the individual cells are not visible to the naked eye.

The sweet taste in algae probably attracted early *Homo*. Our ancestors drank from the top of the water column, drinking directly with their mouth or by using their hand as a scoop. Algae grow and concentrate on the top of the water column, which made drinking an effective method of harvesting algae.

Algae in drinking water would have acted as a natural food supplement to supply the essential nutrients, vitamins and antioxidants, especially the omega-3 fatty acids that provided the green spark for encephalation.[88] Larger brains differentiated our ancestors from their cousins and enabled higher cognitive skills that aided survival.

On the Rift Valley Lakes where early Homo developed larger brains, algae grow to such concentrations that the thick green biomass forms mats on the top of the water column. The wind aggregates algae naturally on the lee side of lakes. These concentrated algae mats would have been easy to gather for supper or store for later use.

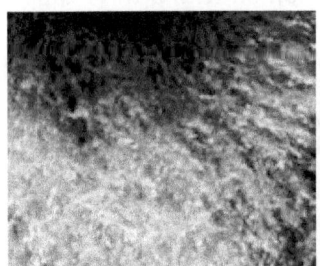

Algae Mats

Spirulina is the bestselling algae nutritive supplement on the market today because it provides a complete set of essential nutrients, including vitamins, minerals, trace elements and antioxidants. The Kanenbu tribe has harvested spirulina for centuries from Lake Chad.

Kanenbu Spirulina Ladies harvesting from Lake Boudou Andja, Chad

Among the many edible algae that would have been plentiful for early hominids in and around Africa were arame, alaria, seawhip kelp, chlorella, dulse, hijiki, karengo, kombu, nostoc, nori, ogo, sea lettuce, sea palm, spirulina and wakame. Algae components would have been available also including agar, alginates and carrageen as well as carotenoids (pigments) such as luten, phycocyanin, zeaxanthin, astaxanthin and phycobiliproteins.

As brains became larger, hominids probably took another small step and exploited the lacustrine ecosystem. The edges of rivers and lakes are filled with nutralent rich aquatic plants, fish, crustaceans, amphibians and birds. Many of these creatures are what they eat and simply concentrate the nutrients from their predominately algae diet. Fish and other aquatic creatures do not synthesize omega-3 fatty acids, they concentrate omega 3 from its source – algae.

Harvesting algae and the meat of algae feeders from rivers and lakes would have allowed early Homo to grow a larger brain and body. Once hominids had sufficient strength and stamina, they would have been ready to take a big step up the food chain to the savanna where they could hunt meat with their strong brains, eyes and bodies.

Color plays a sizable role in motivating appetite. Most seeds, grains and roots are a dull brown and neither pleasing nor appetizing. Algae have colorful pigments the plant uses to collect solar energy and drive photosynthesis. These pigments provide a spectrum of natural colorings to foods. Unlike modern synthetic colorings, algae pigments are not only colorful but also nutritious. Sea vegetables or local freshwater algae would have added attractive color when consumed on their own or when blended with other foods.

Algae were probably our first food because algae were the best tasting, best texture and most colorful food available. Algae were the most plentiful when humans began gathering food. Algae delivered the highest nutralence of any available food source. Most edible algae provide significant protein and a rich assortment of macronutrients as well as essential vitamins and minerals; especially phosphorus, potassium, iron, selenium, copper and zinc.

Taste bud evidence

The human tongue has specialized taste bud receptors for a fifth taste – umami. Umami is a rich flavor constituent found in some protein-rich foods. The unique umami taste which means "good flavor" in Japanese has been isolated and marketed as monosodium glutamate (MSG). Umami translates to brothy, meaty or savory. The unique taste is induced by three proteinogenic amino acids: glutamic, inosinic and guanylic. The savory taste would have been largely absent from early hominid diets and would have been very attractive.

Umami

Glutamate plays a key molecule in human cellular metabolism. Proteins are broken down by digestion into amino acids, which serve as metabolic fuel for other functional roles in the body. Glutamate is the abundant excitatory neurotransmitter in the vertebrate nervous system and regulates several brain functions.

Glutamate's role in body functions is so critical that some animals, including humans, evolved a special taste bud call the mGluR4 receptor to taste glutamate. This umami taste is most concentrated in high protein algae and algae feeders such as fin and shellfish, which would have made these foods quite attractive to early Homo. (Today the umami taste can also be found in milk, aged cheese and some meat products.) The combination of sweet and umami tastes that were not available to our ancestors from terrestrial foods would have provided a strong incentive for early Homo to eat algae.

Human migration

As early humans migrated out of Africa 70,000 years ago, they followed coastlines where macroalgae – seaweeds and sea vegetables – were plentiful. Anthropological records show caves used during human migration hold traces of sea vegetables and algae cuds. Algae cuds come from seaweeds that are chewed like gum to release the rich nutrients bound in algae cell walls.

At low tide, hominids could harvest algae easily and dry it quickly in the sun. The light produce probably served as the first wampum in trade because it was easily transportable. Algae wampum offered a side benefit; a hungry trader could eat the product.

Algae manage insulin release in the blood stream and create a feeling of satiety or fullness that would have been extremely valuable for hominids. Mothers had to carry their children and babies crying from hunger pangs would have created a predator threat to the mother and the tribe. Feeding infants a little algae would have helped brain, heart and eye development from omega-3 fatty acids as well as given them a comfortable feeling of fullness so they did not cry.

Convenience food?

Algae probably provided our ancestors with the original convenience food. Algae can be eaten fresh or dried and stored for up to two years. Dried algae turns dark blue, green or black.

When reconstituted in water, algae recover the similar bright red, yellow, green or purple color it had when it was fresh.

Terrestrial foods were dry, hard, bitter and starchy. Land plants were difficult and risky to gather due to stealth predators. Algae offered a fresh, soft, delicious taste and were easily accessible. In many locations, algae were harvestable year round, which would have been a tremendous advantage when terrestrial crops were dormant or not producing.

Early humans probably rubbed algae oil on their skin for sun protection. Algae add moisture and speeds the recovery from wounds, burns and bruises. Algae's high antioxidant activity protects skin from inflammatory reactions and sun damage.[89] Pacific Rim societies have been using algae for natural foods and remedies for centuries because they are effective. Organic chemists, medicinal chemists, biologists, and pharmacists are currently developing new anti-inflammatory and anti-cancer medicines from algae.[90]

The first written mention of algae was in Korea, about 57 BC. The Samgukyoosa contains passages that record gim, (today called nori) as part of the dowry for Shilla royalty. Members of the Chinese Court, around 1,100, harvested and reserved a specific algae variety for the Chinese Emperor. The Japanese reserved another algae variety for the Samurai, the Japanese nation's fiercest warriors.

The Aztecs used algae for food, medicine, trade and religious ceremonies. Indigenous people along coastlines or lakes have harvested natural stands of algae for millennia for use as food, feed, medicines and trade. Algae probably protected our ancestors against many diseases that plague developing countries today. Algae protects against scurvy, xerophthalmia (blindness from vitamin A deficiency), goiter, arthritis, diabetes, mental retardation and others. The Chinese have used algae for medical purposes for centuries because these natural remedies are safe and effective.

Roman farmers and soldiers valued seaweed as an animal feed supplement. Roman military officers fed algae to their horses to improve color and sheen to their coats as well as health and stamina. Farmers and gardeners used seaweed as a soil amendment to improve the color, taste and texture of produce. The classic red color of Roman military tunics came from pigments extracted from an algae-lichen crust known as urchilles. Wealthy Roman women used the pigments on their cheeks as rouge.

Chinese Olympic athletes have consumed algae for decades because, like the Samurai, algae nutrients enable them to train harder and longer. The therapeutic elements provided by algae allow the athletes to recover from injuries faster. The Chinese have a secret algae cultivation center near their Olympic training center that enables the athletes to eat the algae fresh. Fresh algae maximize the availability of micronutrients, especially antioxidants.

U.S. Olympian Lee Evans won four gold medals at the 1968 games in Mexico and held of four world records in track and field. Evans ate spirulina because the algae improved his training and improved his speed and endurance. Geronimos Dimitrelos was competing for a place on the U.S. Olympics team in Greco Roman wrestling when he broke his wrist severely. He was so impressed with how algae speeded his recovery, that he started a new algae company, AlgaetoOmega to grow algae nutritional products.

NASA recommends algae foods for space flights because research shows algae improve mental acuity, eye sight, digestion and stamina. Algae build the immune system and moderates inflammation. Algae provide another benefit for space flight, habitat renewal. Algae can absorb all the exhaust gasses produced inside the living area and return pure oxygen to the air. Algae can pull all the organic wastes from the liquid and solid waste streams and repurpose them in food, feed, fertilizer or medicines.[91]

Spirulina drink

Spiralps® (spiralps.ch) is a new type of natural soft drink containing fresh spirulina, organic fruits and alpine herbs. The fresh *spirulina* in Spiralps, introduced in Europe in 2012, tastes great, better than dried algae and is nutritionally superior. Processing uses high-pressure pasteurization system that avoids heating the spirulina.

Peace Microfarms

Spiralps® and Controlled Environment Growing System

The next section examines the nature of this exceptional plant.

Chapter 6. What makes Algae Special?

Algae, nature's oldest and possibly best food.

The foundation for freedom foods and abundance growing methods relies on a plant that used 3.7 billion years of evolution wisely to develop strategies to:

- Adapt to acute searing and freezing temperature spikes.
- Survive the extremely hot temperatures of early Earth.
- Grow in ocean, brine, saline, waste and fresh water.
- Live through brutal electrical, wind and ice storms.
- Use either organic material or sunshine for energy
- Go dormant when conditions degrade, yet survive.
- Grow rapidly at any latitude, longitude or altitude.
- Grow faster than any other plant on Earth.
- Thrive using no extracted fossil resources.
- Maximize productivity per unit of space.

All living organisms evolved from algae so it should be no surprise that algae contain all the essential nutrients for life and vitality. A few grams of algae a day act as a nutritional supplement, providing the essential nutrients, vitamins, minerals and antioxidants.

Algae evolution

The first alga (singular) cell was among the earliest life forms on Earth 3.7 billion years ago. The prokaryotic cells of blue-green algae, cyanobacteria, were elementary and contained no nucleus or other membrane-bound organelles.[92] These organisms lived in oceanic ecosystems synthesized by abiotic, high-energy processes including lightning, ultraviolet radiation and pressure shock. The atmosphere was anaerobic with high levels of methane and ammonia but no oxygen. These plants developed photosynthesis and produced oxygen in our atmosphere, which enabled higher life forms to evolve.

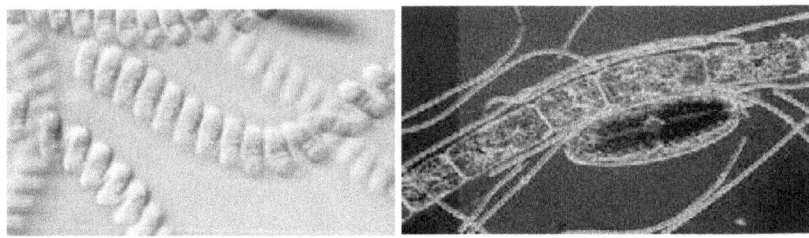

Cyanobacteria – blue-green algae with no cell wall

Eukaryotic green algae (Greek for "true nut") plants have cells with their genetic material organized in organelles. They evolved 2.7 billion years ago and have discrete structures with specific functions. These plants have a double membrane-bound nucleus or nuclei. The basic single-celled organism, algae, has the general appearance illustrated in Figure 6.1. The University of Montreal, U.C. Berkeley, University of Texas and others host culture collections of algae species with descriptive details and pictures.[93]

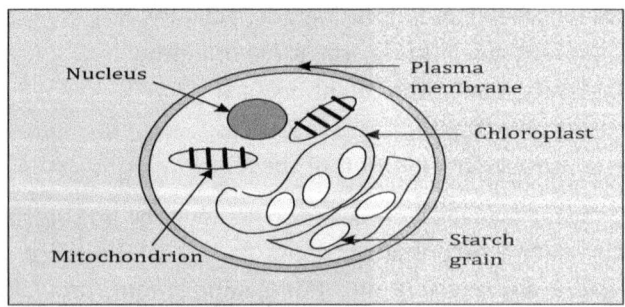

Figure 6.1. Alga Cell

What makes Algae Special?

Major steps in cell complexity occurred with the evolutionary progression from single cell prokaryotic cells to multicellular eukaryotic cells. The addition of a nucleus and cell wall were huge evolutionary advances. Cell walls enable algae to protect itself from the surrounding environment, typically water and pressure, called osmotic pressure. Cell walls regulate osmotic pressure produced by water trying to flow in or out of the cell through its semi-permeable membranes due to a differential in the solution concentrations. Algae typically possess cell walls constructed of cellulose, glycoproteins and polysaccharides. Some have a cell wall composed of silicic (silicon) or alginic acid.

Many species are single-celled and microscopic including phytoplankton and other microalgae. Others are multicellular and may grow large such as kelp and Sargassum. Phycology, the study of algae, includes the study of prokaryotic and eukaryotic species. Some algae live in symbiosis with lichens, corals and sponges.

Other species are made of fine filaments with cells joined from end to end. Some clump together to form colonies while others float independently. Seaweeds may grow in nearly any shape such as cones, tubes, filaments, circles or they may imitate the shape of land plants. Seaweeds developed in parallel evolution with land plants.

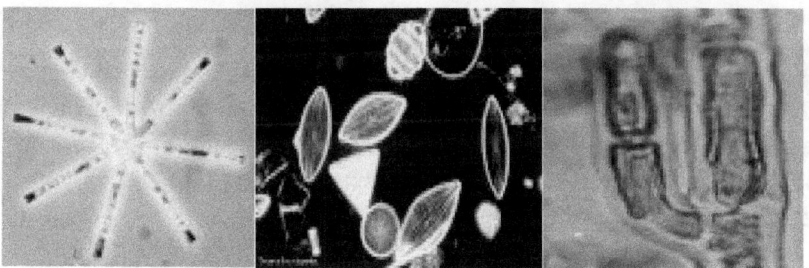

Algae Cell Walls

Algae can be lively little critters even though they are not animals. Many can swim, such as dinoflagellates that have little whip-like structures called flagella. Some use the flagella to pull or push themselves through the water. Some algae squish part of their body forwards and crawl along solid surfaces.

Algae, often called microscopic phytoplankton, grow in most bodies of water, moist places, on and in trees, and even in rocks. This little plant provides the foundation for the food chain, feeding both microbial and animal plankton, zooplankton and fish. Subtract algae and phytoplankton from the water column and fish, shellfish, reptiles and other aquatic creatures cannot survive.

Since algae form the bottom of the food chain, everything around acts as a predator. Algae's strategy to predation is brilliant – grow faster than consumers can eat. A single alga cell may produce one million offspring in a day. At night, algae take a well-deserved rest in a phase called respiration. While an individual alga cell is not visible, algae communities appear first as a cloud and then as tiny specks that are cell clusters. Some algae aggregate to form structures, such as filaments, globes, wheels or with spirulina, spirals.

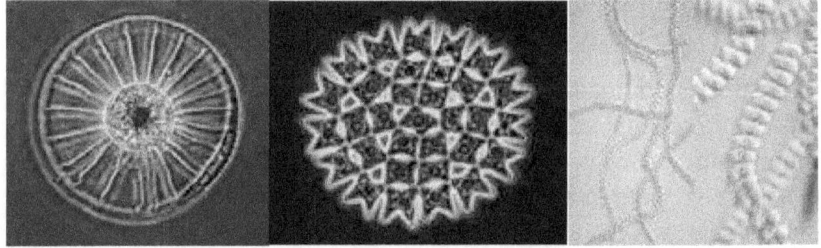

Microalgae

Algae break the rules for plant classification because they evolved in many different forms – cells, multicellular plants, bacteria and in nearly infinite combinations. While the various species share certain characteristics, different algae display extraordinary variety in shape, size, structure, composition and color.

Classifications

The major groups of algae have been distinguished traditionally based on pigmentation, shape, structure, cell wall composition, flagella characteristics, and products stored. Algae display so many variations they express exceptions to nearly every classification rule.

What makes Algae Special?

Diatoms, stoneworts and dinoflagellates

The broad algae classification includes:

- Bacillariophyta – diatoms
- Charophyta – stoneworts
- Chlorophyta – green algae
- Chrysophyta – golden algae
- Cyanobacteria – blue-green
- Dinophyta – dinoflagellates
- Phaeophyta – brown algae
- Rhodophyta – red algae

Each species has its unique constituencies and produces various levels of each compound during its life cycle. Producers may stress algae by withholding a specific nutrient or changing the temperature or pH to prompt algae to overproduce the target compound such as lipids.

Red algae are a large group of about 10,000 species of mostly multicellular, marine algae, including seaweed. These include coralline algae, which live symbiotically with corals, secrete calcium carbonate and play a major role in building coral reefs. Red algae such as dulse *(Palmaria palmata)* and laver (nori/gim) are a traditional part of European and Asian cuisine. Red algae are also used to make many other products such as agar, carrageenans and other food additives.

Productivity

Scientists have known algae's food value for centuries and food potential for at least 100 years. Consider the annual protein production per acre for food grains calculated using half its theoretical photosynthetic capacity, Figure 6.2. Algae provide a superior set of vitamins and minerals than found in land plants. Algae are not a full solution for malnutrition because the biomass is light on calories. Fortunately, calories are cheap and easy to add to a diet.

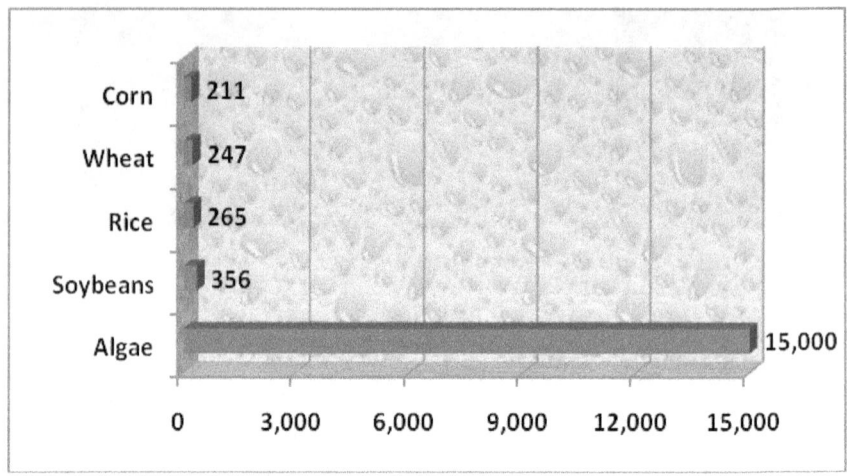

*Figure 6.2 Algae Protein Production Potential –
Pounds per Acre per Year*

Land plants have specialized cells for moving nutrients and for reproduction that algae do not need. Algae are distinguished from the higher plants by a lack of true roots, stems, or leaves. Some seaweed appears to have leaves or trunk but they are pseudo leaves made up of the same cellular structure as the rest of the plant.

Algae grow much faster than land plants because they do not have to invest energy in specialized structures such as roots, stems, leaves or reproductive appliances. Land crops have to produce cells specialized for each separate function, such as reproduction or nutrient transport. Algae grow as single cells without specialty functions, which saves considerable energy and allows each cell to reproduce rapidly.

Algae use nitrogen to manufacture amino acids, nucleic acids, chlorophyll and other nitrogen compounds. Green algae absorb available nitrogen from their culture. Many cyanobacteria species are able to fix nitrogen absorbed from the air, as well as from water, in a process known as diazotrophy. Since the atmosphere is 78% nitrogen, nitrogen fixing is a strong competitive advantage for growth because water-based nitrogen is often limited.

What makes Algae Special?

These algae fix nitrogen naturally, without added energy. When used with irrigation, the algae attach to plant roots and pull nitrogen from the atmosphere, which is absorbed by crops. About 90% of the cost of commercial nitrogen fertilizer comes from the energy, typically natural gas, used to extract nitrogen from the air.

Variation

Algae range from microscopic single-celled organisms to multicelled organisms. These plants thrive all over the world in marine and fresh water environments – nearly any moist environment. Terrestrial algae grow in all types of soils where they can capture nitrogen from the air that can be used through the roots of plants. They may be free-living or live in symbiotic association with a variety of other organisms such as bacteria, lichens and corals.

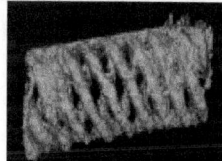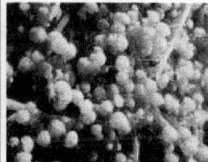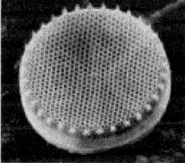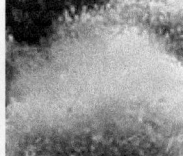

Algae Shapes

Each species may exhibit multiple strains with unique characteristics. A single strain may display completely different structural expression and composition in different growing conditions with variations in light, temperature, nutrients, mixing or water pH.[94]

Macroalgae

About 10% of algae species are macroalgae, (seaweed and sea vegetables) such as kelp that can grow to 60 meters. Most are microalgae that occur in every color, shape and small size imaginable.

Away from the oceans, most algae grow in, on or among the roots, trunk and leaves of land plants. A handful of local dirt may hold several billion algae cells and over 100 algae species. Land plants need algae to break down chemical fertilizers so they are bioavailable and absorbable by the plant roots. Land plants and algae work symbiotically as algae supplies nutrients and plants provide a protected area in which to grow. Algae also support symbiotic relationships with mosses, fungi, yeasts, lichen, corals and sponges.

Macroalgae – Seaweeds and Sea Vegetables

The next chapter introduces algae cultivation in peace microfarms.

Chapter 7. What are Peace Microfarms?

Envision 10 million Green Masterminds growing food and other valuable coproducts in peace microfarms globally. Imagine what each grower can do for the health and vitality of her family and community.

Freedom foods grow in peace microfarms that save precious natural resources for our future generations. Peace microfarms offer a novel approach to avoid conflicts over the fossil resources currently used to produce food. Planet Enriching Algae Cultivation Ecosystems, (PEACE) microfarms use abundance methods that enable growers distributed globally to recycle nutrients and energy from sterilized waste streams.

Many people today cannot grow food locally because they lack the weather, fossil resources or money required for crop inputs. When fully developed, peace microfarms will allow growers and communities globally to use affordable inputs to grow foods locally. Peace microfarms grow food and other forms of energy sustainably independent of climate, altitude, latitude, geography or politics.[95]

Peace microfarms are under development and we need your help. Please follow R&D progress and contribute your ideas for designing effective microfarms at www.AlgaeCompetition.com

Peace Microfarms

Peace microfarms are adaptable microcrop platforms that enable growers to use low cost inputs to cultivate a wide variety of high value products. Microfarmers practice abundance and use green solar, sunshine, for energy. They may recycle organic inputs from farms, gardens, kitchens or other waste streams that are surplus, low-cost or free. Growers cultivate microorganisms such as algae and the microflora algae attract to produce food for people and feed for fish, fowl, dairy, and meat animals. Other growers grow and flow their culture to produce rich organic biofertilizer for gardens or fields.

Growers practicing abundance are essentially green solar gardeners as they transform solar energy to rich, nutritious plant biomass. The green biomass concentrates energy in chemical bonds that are portable and may be used directly for food or transformed to many other forms of energy. Microfarms do not need genetically engineered seeds because microcrops offer substantial natural biodiversity. Microfarmers use four configurations to grow protein at much faster rates than land crops, Table 7.1.

Microfarm configuration	Estimated protein yield compared with corn
1. Open pond or raceway	20 times
2. Covered pond or raceway	25 times
3. Semi-closed culture	30 times
4. Closed or controlled environment	Over 30 times

Table 7.1. *Microfarm Configurations*

Covered microfarms allow growers to extend the season by two to four months. Semi-closed and closed systems allow year-round production, independent of weather. Microfarmers train indigenous, local algae to produce proteins, oils, carbohydrates and other coproducts rapidly. Some growers cultivate exotic species from algae collections but locally adapted species typically outperform others. Figure 7.1 shows how peace microfarms transform low cost inputs into a broad spectrum of high value products.

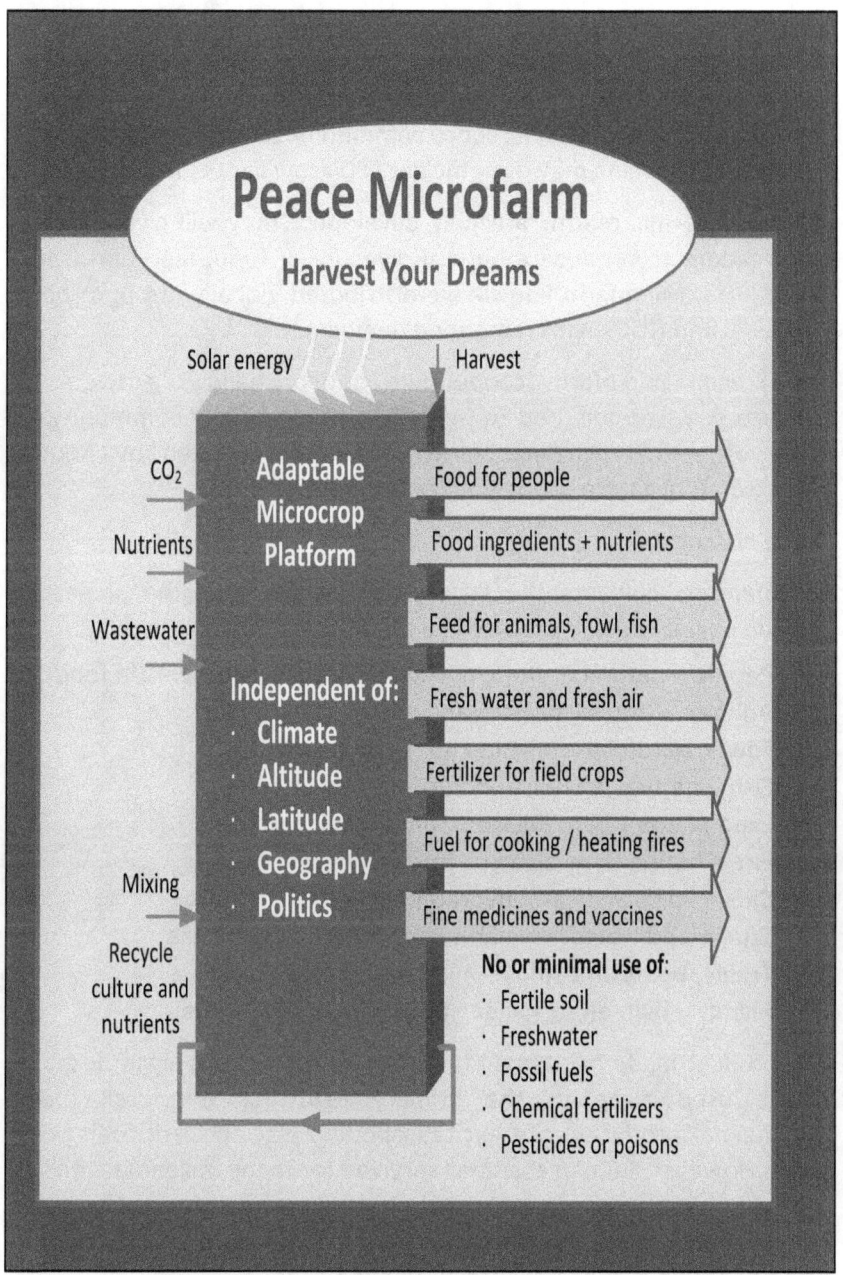

Figure 7.1. *Peace Microfarm Features and Products*

Microfarms can be sited in nearly any location. The footprint may fit in a corner of a backyard, rooftop, balcony, barn, field, wetland, desert, prairie or other non-crop land. Microfarms may serve a family, community garden, village, coop, community or city. Locations with insufficient sunshine may use efficient LED grow lights for energy.

When peace microfarms are fully developed, they will give growers the freedom to produce a natural diversity of food, feed and other coproducts. Microfarm knowledge distributed globally through social networks and NGOs will create food democracy.

The French microfarm cooperative network allows growers to produce spirulina for food to benefit their family and community. A video of the French microfarm cooperative created by Robert Henrikson is available at www.AlgaeCompetition.com.

Algae energy

Microfarmers cultivate the fastest-growing plant on the planet to provide portable energy usable in a multitude of ways, including:

- **People** – natural protein, nutrients and micronutrients in food.
- **Animals** – natural protein and nutrients in fodder.
- **Fowl** – natural protein and nutrients for birds.
- **Fish** – natural protein and nutrients in fish feed.
- **Land plants** – rich, full spectrum organic fertilizer.
- **Fire** – high-energy algae oil for cooking and heating.
- **Cars** – lipids and carbohydrates refined to biofuels.
- **Trucks and tractors** – high-energy clean, green diesel.
- **Trains, boats and ships** – high-energy clean diesel.
- **Planes** – high-energy, clean aviation gas and jet fuel.

Any product made from fossil fuels can be made from algae because nature used algae as the primary feedstock for fossil fuels. Commercial producers are excited about replacing fossil fuels with algae. However, human societies survived for many millennia without the convenience energy sources derived from fossil fuels. The most critical energy source for humans is food. We survive only a short time when deprived of the vital energy supplied by food.

What are Peace Microfarms?

Growers cultivate microbial communities, which may be pure strains of algae, but are often diverse communities of algae and the multitude other microorganisms algae attract. Microflora communities thrive in aquatic and moist terrestrial settings and include algae, fungi, bacteria, viruses, slimes and other tiny organisms. This diverse array of microorganisms works symbiotically to produce compounds valuable to plants, animals, fish and humans.

Microfarms can produce cosmeceuticals, nutraceuticals, vitamins, pharmaceuticals, vaccines and medicines. Of course, higher value products require scaled microfarms and sophisticated extraction technologies. Innovations in growing, harvesting and extraction methods will make high value products more practical for microfarms.

Microfarms are not designed to scale for biofuels, although biological or technical breakthroughs may make biofuels practical. In many part of the world, cooking and heating oil has a high value. Black smoke death from wood, coal and dung cooking fires claims over 5 million women and children a year and leaves tens of thousands with respiratory diseases. Microfarmers can grow algae and press out the oil for cooking and heating and feed the residue to their animals. Algae oil burns cleanly, without black particulates because it is simply vegetable oil.

Industrial or organic farmers may use microfarms to recover and reuse the energy and nutrients in the farm waste stream to reduce production costs while improving soil fertility and crop yields. Urban gardeners may source nutrients from municipal and industrial waste streams to grow rich algae biofertilizers that speed plant growth and development as well as increase produce size, taste, texture, color, nutrition and quality.

Microfarms provide substantial value for traditional farmers by avoiding fossil inputs, increasing productivity, and reducing costs. Microfarmers can improve soils and reduce water, energy and fertilizer waste while decreasing soil erosion and air, water, and soil pollution, Table 7.2.

Benefit	Crop inputs and costs
Fossil inputs	Eliminate or minimize scarce and expensive fossil inputs including fertile soil, fresh water, fossil fuels, fertilizers, and fossil agricultural chemicals.
Plentiful inputs	Produce food, coproducts and other forms of energy using solar energy, CO_2, and wastewater.
Improve yields	Enhance yields of protein, lipids, carbohydrates, energy and other target compounds 30 times.
Crop diversity and nutrition	Expand crop diversity, providing better nutrition, micronutrients, vitamins, and minerals.
Embrace global warming and climate chaos	Crops produce effectively despite prolonged heat, droughts, more extreme storms, salt invasion, rising oceans, wild fires, and pests.
Enable the poor and malnourished	All people can produce food when the inputs are free or surplus, assuming they have access to growing systems and sufficient training.
	Boost yields for field crop farmers
Texture and taste	Improve produce texture and taste through the immediate bioavailability of micronutrients.
Productivity	Improve crop yield, speed to maturity, size, weight and quality 30-50% by providing immediately bioavailable nutrients.
Vitamins and minerals	Enhance the presence, quality and availability of vitamins and minerals 20-30% in produce with bioavailable nutrient and micronutrient delivery.
Reduce costs	Transform a cost, waste management to a profit center that reduces crop input costs.

Regenerate soils	
Soil compaction	Reduce soil compaction and increase porosity 500% to stimulate root growth, make room for microflora and worms to enhance plant strength.
Crust	Strengthen the soil crust to add nutrients, organic material, and minimize erosion.
Soil structure	Improve topsoil structure by expanding the humus and organic material in the soil.
Soil microbes	Use algae to attract microbial communities that act to enhance crop health and productivity.
Soil moisture retention	Improve soil moisture retention and decrease heat and drought stress.
Improve agroecology	
Fertilizer pollution	Reduce air, soil and water pollution by using fewer chemical fertilizers.
Erosion	Minimize soil loss to wind and water.
Agricultural chemicals	Minimize pollution from agricultural poisons by diminishing or eliminating them.
Bioavailable nutrients	Deliver bioavailable nutrients to the soil precisely when plants most need them.
Greenhouse gases	Reduce greenhouse gas emissions, especially CO_2, methane and nitrogen oxides.
Tillage	Reduce the need for tillage and soil disruption.
Organic farming	Support and accelerate the transformation from industrial farming to organic farming.

Table 7.2. *Peace Microfarm Benefits for Farmers*

Microfarmers cultivate algae and possibly other microorganisms as they follow one or a combination of four Sustainable and Affordable Food and Energy, (SAFE) production paths, Figure 7.2. Peace microfarms support all four pathways.

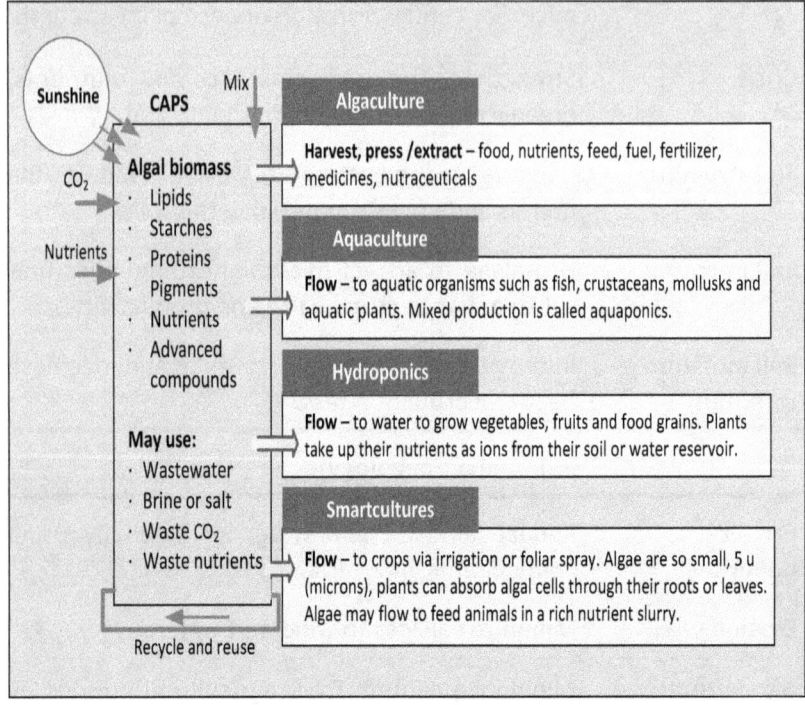

Figure 7.2. *Abundance SAFE Production Paths*

Algaculture

Algaculture grows microalgae or macroalgae, (seaweed) for commercial purposes or domestic needs. Extraction of the algae biomass enables the farmer to use the energy, nutrients and a wide spectrum of valuable coproducts, Table 7.2. About 33% of the algae grown commercially today feeds fish and shellfish. Nori, used to wrap sushi, leads the sea vegetable market and exceeds $3 billion globally. Spirulina leads the microalgae market with about 5000 metrics tons a year. Most spirulina is currently sold as a health supplement to provide essential micronutrients.

Food	Biofuels	Novel Solutions
Primary • Protein • Lipids – oils • Carbohydrates • Nucleic acids	**Primary** • Gasoline • Clean diesel • Methanol/ethanol • JP-8 jet fuel	**Air** • Carbon sequestration • Carbon capture/recycle • Capture sulfur • Capture heavy metals
Secondary • Flour • Meat enhancer • Ice cream • Milk substitute • Sugar substitute • Sea vegetables • Food ingredients • Emulsifiers and thickeners • Novel flavors and textures • Pigments • Health foods • Nutraceuticals • Omega 3s	**Secondary** • Aviation gasoline • Alcohols • Hydrogen • Asphalt • Plastics, biodegradable • Rubber imitation **Biofertilizers** • Organic N-P-K • Bioavailable target nutrients • Micronutrients • Plant hormones • Soil organics	**Water – clean** • Waste streams – municipal, industrial, farm, brine and ocean • Recover heavy metals **Cosmetics** • Moisturizers • Skin care **Local algae production** • Foreign aid • Disaster relief • Hunger and poverty **Medicines** • HIV / AIDS and SARS • Vaccines
Feed and fodder • Pets, fish, fowl • Meat animals • Micronutrients • Medicines	• Build soil structure • Improve porosity • Plant growth regulators • Natural pesticides • Natural herbicides	• Antibiotics /antiviral • Burns and bruises • Stomach remedies • Anti-cancer toxins • Pharmaceuticals • Advanced compounds

Table 7.2 *Algae for Food, Biofuels and Novel Solutions*

Algaculture producers use many microfarm shapes, sizes and forms including ponds, troughs, semi-closed and closed systems. Food, health food, feed and nutraceutical producers include Earthrise, Cyanotech, Algae Biosciences, Parry Nutraceuticals, Solazyme, Seambiotic, Cellana, Martek Biosciences and Boonsom, Thailand.

Microfarms sited near a carbon source such as a waste pile on a farm, coal fired power, cement or manufacturing plant gain the advantage of a free carbon source. Every ton of algae consumes nearly two tons of CO_2, so free carbon reduces operational costs.

Other microfarms are sited near a wastewater treatment facility to gain access to free nutrients. Some growers may source organic wastes from farm, garden or other waste streams. Algae grow well in fresh water but communities have competing needs for sweet water. Many communities have substantial sources of gray brackish or wastewater that is not potable but excellent for growing algae. Some farms have reservoirs, ponds or wetlands that capture farm runoff, which are perfect for growing algae.

Grow and flow – hydroponics   *Grow algae to feed vegetables*

Half of the water stored in the earth's crust is brine water, which is too salty for human use or for irrigation. Algae thrive on brine water, which often carries the full spectrum of essential nutrients. Many deserts, including in the U.S. Southwest, have huge underlying oceans of brine water in relatively shallow aquifers. These brine aquifers could produce millions of tons of algae biomass for 400 years.

Growing algae as fodder for animals, birds or aquatic creatures will be popular in many settings because animal feed requires lower levels of cleanliness, (except for pets in the U.S.), than producing food for direct human consumption. Pet and animal foods that enter the human food chain have high quality requirements, similar to human foods.

Wastewater microfarms can produce food quality algae with the proper safeguards in place. Many human-grade valuable coproducts may be extracted from wastewater algae such as vitamins, minerals, antioxidants, trace elements pigments, oil and carrageen.

Hydroponics

Algaculture producers may produce a slurry or solid product similar to fish fertilizer for use in hydroponics. Farmers can grow algae next to their hydroponics unit and a portion of the algaculture flow to containers where vegetables, grains and fruits grow in the rich algae water. Algae provide all of the macro and micronutrients necessary to grow large, colorful and tasty produce.

Hydroponic farmers grow plants using mineral solutions in water rather than soil. In natural conditions, soil acts as a mineral nutrient reservoir but the soil itself is not essential for plant growth. Terrestrial plants grow well with their roots in an inert medium such as perlite, gravel, mineral wool or nutrient solution. Hydroponic crop yields are be no better than crop field yields with good soil. Crop yields are limited by factors other than mineral nutrients; especially light. Hydroponics growers produced vegetables on Pacific volcanic islands that lacked fertile soil in World War II. Hydroponics saved considerable transportation cost during the war. Similar growing systems are used today to feed scientists in Antarctica.

Aeroponics, developed largely by NASA for space travel, grows plants in an air or fine mist environment without soil or aggregate medium. Aeroponics culture differs from both hydroponics and *in-vitro* production (plant tissue culture). Unlike hydroponics that uses water as growing medium and essential minerals to sustain plant growth, aeroponics cultures grow without an aggregate medium.

Growers transmit nutrients by water mist, so aeroponics is actually a form of hydroponics.

Aeroponics – Fine mist systems

Aquaculture and Aquaponics

Aquaculture farmers grow fish and shellfish that feed on aquatic plants such as algae. Algae represent the preferred diet for most fish fry (immature fish) because the cells are small enough for the fry to eat. Most fish evolved on an algae diet in their natural settings. Most fish grow faster and have fewer digestive problems on algae compared with food grains.

The Chinese have practiced aquaculture since 2500 BC. Today, half the world's commercial fish and shellfish production comes from aquaculture. A recent scientific study reported that over 90% of the large fish have been extracted from the oceans. Unfortunately, fishermen overharvest many of the smaller fish too, depleting the food chain. With diminishing natural fisheries in oceans, rivers, lakes and estuaries, aquaculture will play a larger role in our food supply.

Aquaponics integrates fish and plant farming. Farmers grow algae to feed fish that add urea to the water. The nitrogen rich water flows to hydroponic greenhouses where vegetables and fruits grow in the high nutrient water. Polycultures can grow food with renewable energy and in closed systems, minimize consumption of fossil resources, including power and fresh water.

What are Peace Microfarms?

Growing algae to feed fish – Aquaculture

Smartcultures

Sustainable Micro Algae Regenerative Technologies, (smartcultures), enable field crop farmers to recover, recycle and reuse the energy and nutrients in their farm's waste stream to improve crop quality, taste and yields, while reducing operational costs.[96]

Smartculture farmers skip the harvest step and simply "grow and flow" the algae culture directly to their fields to recycle organic fertilizer to their crops that is immediately bioavailable to the plants. Smartcultures deliver 74 nutrients and trace elements that plants use to grow faster and stronger and produce higher yields.

Smartcultures employ a set of technologies that imitate nature to provide enhanced foundation (soil structure) and food (nutrients) to plants. Every farmer and gardener knows plants thrive in amended soils; they grow faster, stronger, and larger, and they have better taste and texture. Animal farmers can recover most of the energy and nutrients remaining in the farm waste stream and recycle it to feed farm, dairy, poultry or meat animals

Smartcultures begin at the crop foundation – soil – with tiny microflora attracted by algae in plant roots that are ingeniously self-regulating and self-regenerative. Smartcultures move farmers toward abundance production by significantly reducing, but not eliminating the use of fossil resources for growing field crops. Farmers using smartcultures are able to leave every field better than they found it.

Field crop farmers using smartcultures can use 80 to 100% fewer agricultural chemicals because algae biofertilizers provide growth hormones that make plants stronger and able to produce natural pest and disease defenses. Growers can reduce soil compaction 500%, enabling significantly longer and stronger root structure. Stronger roots give plants a deeper reach for nutrients and soil moisture. Healthier plants on a stronger foundation need less water and are less vulnerable to weather, winds, weeds, disease and pests.

Algae deliver biofertilizers to field crops through irrigation or spraying. The algae continue to grow in the field, increasing valuable organic material, humus, and improving moisture retention and soil fertility.

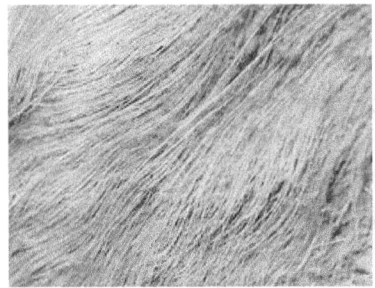 

Algae growing in a Turf Scrubber and Air Dried

Algae can remove nitrogen and phosphorus in livestock manure runoff. Walter Mulbry, an Agricultural Research Service scientist set up four algae turf scrubbers outside dairy barns. The shallow 100-foot raceways are covered with nylon netting that creates a scaffold where the algae grow. For the next three years, from April until December, a submerged water pump at one end of the raceways circulated a mix of fresh water and raw or anaerobically digested dairy manure effluent over the algae. The raceways supported thriving colonies of green filamentous algae.[97] Algae grown in open raceways create diverse microbial communities of microalgae and other microflora.

Each acre of algae raceway removed nutrients from 20 cows' manure. The system recovered 60-90% of the nitrogen and 70-100% of the phosphorus from manure effluent.

The research team calculated the recovery cost was comparable to other manure management practices—with a per pound cost of $5 for N and $25 for P. Dried algae made excellent organic fertilizer as corn and cucumber seedlings grown in algae-amended potting mixes performed as well as those grown with commercial fertilizers.

Field trials

Two years of field research by Mark Edwards with a multinational food production company shows smartcultures methods can:

- Increase income from 20 to 50% by improving crop yield and quality – micronutrients, vitamins, antioxidants, color, taste, texture and shelf life.
- Lower fuel consumption by 20 to 30%.
- Decrease chemical fertilizers by 30 to 50%.
- Reduce air, soil and water pollution by 80 to 95%.

Farms without irrigation systems spray the algae solution on the fields. The algae not only provide an organic fertilizer delivery system directly to the roots of crops but algae continues to grow in the field, as long as moisture is present, regenerating soils and creating additional organic material. Algae's ability to extract in situ nutrients provides a tremendous advantage. Most farmers have available waste streams from human, animal and vegetative wastes on which algae can thrive. Rather than spending 30-40% of their production costs on fertilizer, algae may cut nutrient cost in half because algae can recover 90% of the nutrients from the farm waste stream. The typical farm waste stream contains about half of the nutrients needed for the next crop. Therefore, the net fertilizer reduction approaches 50%.

Smartcultures transform agricultural methods and rather than mining and paying high prices for chemical fertilizers and using them once, farmers can continuously recycle nutrients. Rather than systemically extracting soil nutrients and organics, farmers can cultivate algae and microflora to add nutrients and organics to their fields. Rather than using chemicals that destroy soil microbes and soil structure, smartcultures cultivate microbial communities that improve soil structure. Industrial agriculture degrades soil, promotes erosion and creates severe pollution, while smartcultures improve soil structure and reduce nutrient waste, erosion and pollution.

Smartcultures grow algae in the farm waste stream to fertilize fields.

Recent fertilizer price escalation has pushed the cost of fertilizer up to 35% of farm operating costs. Farmers face two other serious problems with industrial fertilizers: bioavailability and erosion. Microorganisms in the soil, e.g. algae, must first break down chemical fertilizers before the plant can absorb them. The process may take months or years. Consequently, farmers have to put on far more fertilizers than the plant needs in order to maximize growth. Much of the applied fertilizer does not reach the crop and erodes with irrigation, winds and rains. The next year, the farmer must apply even more fertilizer to achieve the same yields. This model is sustainable only as long as fertilizers are cheap and soils do not wear out.

Farmers can improve the quality and quantity of field crops because algae biofertilizers are immediately bioavailable to the plants and create almost no waste. Some algae are able to unlock nutrients in the soil, such as phosphorus. Over 95% of the P in some soils is locked in large molecules that are not absorbable by plants. Algae can solubilize the P and other elements, making them bioavailable.

Smartcultures can deliver precise amounts of target nutrients carried in algae biofertilizers at specific times during a crop's growing cycle – which can maximize germination, early growth, maturation and fruiting. Microfarms near fields can overload one or more nutrients and deliver them to crops exactly when needed. For example, adding more calcium when the crop is fruiting may enhance fruit size, weight, color, texture and taste.

Farmers can save money and energy by lowering their consumption of fuel due to easier cultivation. In some settings, smartcultures have improved soil porosity (loosen compacted soil) by. Smartcultures reduce air, soil, and water and pollution because algae biofertilizers and plant growth hormones significantly diminish the need for agricultural chemicals.

Drip irrigation can deliver algae biofertilizers precisely to the roots, minimizing water use and the waste of nutrients. Algae continue to grow in the soil while moisture is present, which adds rich organic matter and conditions the soil, making it more erosion resistant. This model may also use no or minimal till to minimize soil disruption and provide longevity to the water-efficient drip irrigation system. In non-irrigated settings, growers may apply the algae culture with a field or aerial sprayer.

The seed catalogue, *Cook's Garden* sells a seaweed product called Sea Magic.[98] The algae fertilizer contains cytokinins and 17 amino acids that encourage stronger, lusher growth, more sugar production in fruiting crops, and increased blooms. The company claims this seaweed fertilizer increases yields, performance and flavor. Our field and garden research supports their claims, which include 24% more tomatoes, 25% more grapes, 34% more cucumbers, and 47% more peppers.[99] Our peppers were closer to 25% yield improvement, but our garden already benefits from excellent organic soil with microorganism communities. Most nurseries and seed catalogues sell multiple algae fertilizer products made from mined fossilized seaweed. Similar seaweed fertilizers are plentiful in hydroponic supply stores and claim similar yield boost.

Other forms of food production

Human societies will need every sustainable form of food production to meet global food demand, including organic production, controlled environment agriculture and vertical farming.

Organic production offers health advantages for consumers and producers such as the avoidance of pesticides. Organic production is not sustainable because it consumes more cropland, water and fuel than industrial agriculture. Organic produce spoils faster than industrial foods, which creates more waste. Organic growers use less inorganic fertilizer and agricultural chemicals, which creates less pollution.

Most countries, including the U.S. have insufficient cropland to grow the biomass needed for organic compost for fertilizer. Even if sufficient compost was available, the energy and time costs involved in harvesting, transporting, storing, turning and cultivating the material into fields preclude widespread adoption.

Controlled environment agriculture and vertical farms offer the promise of resource efficiency and pollution free food. Capital and operating costs may be too high for widespread use, unless policy leaders tax food distribution miles or agricultural pollution. Vertical farms have been proposed for decades but production costs undermine profitability and viability. As the costs of fossil resources increase and severe weather causes more crop losses, vertical farms will become more attractive.

Chapter 8. Microfarm Examples

*God Almighty first planted a garden. And indeed, it is
the purest of human pleasures.* **– Sir Francis Bacon**

A few small-scale algae cultivation systems are available currently.
Several companies will provide scalable algae cultivation kits in the
near future. Current systems require considerable attention,
knowledge and labor. The best solution will include growing support
that minimizes the learning curve and maximizes grower success.

Home growers have not cultivated algae successfully, except for
spirulina. Cultivating algae successfully today requires considerable
knowledge, experience and technology. The industry needs a solution
that empowers gardeners and farmers to grow algae on a small scale
and then migrate to larger cultivation systems.

The key innovation will be the addition of remote monitoring
capability to growing systems. Culture information will upload
through the Internet to a laboratory staffed with experts that provide
advice to growers. Eventually, expert systems will provide grower
advice in real time, which will greatly simplify algae cultivation.

About 150 commercial algae producers cultivate algae in 2012, primarily for special compounds such as omega-3 fatty acids, nutraceuticals, and health foods. Firm profiles are available at the Algae Biomass Organization and *Algae Industry Magazine*.

Home systems

Several providers offer algae kits that enable home growers to produce small amounts of algae.

AlgaeLab.org

Aaron Wolf Braun founded algaelab.org and teaches workshops on how to grow algae. Participants complete the seminar with a take-home growing kit similar to the picture. Aaron, a.k.a. Dr. Friendly, is working on a book on growing algae that includes science trends and technology, and aquaculture from micro to large scale.

The AlgaeLab store offers the basic supplies needed to grow spirulina at home. The AlgaeLab.org site has pictures of the cultivation and harvest processes.

Home cultivation systems currently produce primarily the blue-green algae spirulina because it is easiest to grow and harvest. Spirulina has many advantages, including 70% protein, high-density micronutrients, antioxidants, vitamins, minerals and trace elements.

One of the most common requests people make is "How can I grow my own algae?" Jean-Paul Jourdan published manuals and curriculum on how to grow spirulina, encouraging many more people to get personally involved in growing algae locally for family, friends and community. Algae microfarms fit with the growing Do-It-Yourself movement. Microfarms align with the trend for growing food and herbs indoors, in greenhouses, on rooftops and in empty lots urban, backyard and community gardens.

SmartMicrofarms.com

Robert Henrikson, CEO of SmartMicrofarms.com and AlgaeComeptition.com, grows spirulina in several scalable indoor, porch and covered backyard units. He harvests up to a pound daily. His family and friends eat about half fresh. The rest is frozen to cubes or dehydrated.

Development of appropriate scale low cost technology has been ongoing in India by agencies of the government with NGOs. In Africa, humanitarian groups from France have built village projects, including Dr. Ripley Fox and Antenna Technologies.

Appropriate scale village farms in the developing world have led to the emerging movement of commercial spirulina algae microfarms. Growers produce for their families and local markets. Many of the charitable organizations diffusing algae production originated from France. Over 100 algaepreneurs produce algae locally in France today and over 500 are expected within five years. The algae microfarm movement that began in France has recently migrated to Spain and shows signs of coming to North America.

Spirulina producers have formed a cooperative to learn from one another in Southern France to as far north as Normandy. The school at the CFPPA Center in Hyères trains growers. In 2010, growers established the Fédération des Spiruliniers de France and developed a Charter of Good Business practices.

Laurent Lecesve, at his Eco-Domaine farm in Normandy, France.

Cultivation in ponds

Algae pond systems in the U.S. were first developed for wastewater treatment. Producers recovered the biomass, converted it to methane, and burned it as a local source of energy.[100] When fuel was cheap, the energy value from algae was considered incidental.

Wastewater Treatment and an AquaFlow Pond in New Zealand

Algae wastewater treatment offers a low energy and low cost means for cleaning polluted water. Proven technologies kill parasites and pathogens in the wastewater and algae remove the substantial organic material. Producers can recover the algae biomass for use in animal feed, fertilizer, green chemicals and advanced compounds.

Some producers use the algae oil for green energy and the residual protein for animal feed.

A variety of other algae growing systems are available. In open ponds, the productivity ranges from 10-35 grams/meter^2/day or (36-128 metric tons/hectare/year) on a dry weight basis.

Algae Grown in Troughs and Raceways

Algae ponds typically are shallow, six inches to three feet deep, in order to maximize cell access to light. Algae grow quickly and new cells shade older cells. Unmixed ponds have growth only in the top two inches of the water column. Ponds are mixed with a paddle wheel or compressed air that keeps the culture moving around the raceway and up and down the water column. Water movement needs to create sufficient turbulence to move cells to the surface so they can absorb photons. Large ponds used for municipal water remediation typically bubble a mixture of CO_2 and air to move the water.

Size affects water circulation, operating costs, mixing systems and species selection. Mixing gives cells access to light, prevents cells from settling to the bottom and avoids thermal and oxygen stratification in the pond. Effective mixing increases cell density, which reduces harvesting cost. Harvesting typically occurs with microscreen, centrifuge, filters or flocculation.

Raceways scale any size and are constructed as a loop. Raceways have the advantage of simple, low cost construction and maintenance. Most algae production today occurs in open raceways because raceways are cheap to build and operate. Some algae, such as *dunaliella* are grown in deep saline ponds with little mixing. Ponds are most productive in tropical, subtropical and temperate areas with warm temperatures, low rainfall and little cloud cover.

Disadvantages of ponds include lower productivity due to lack of temperature control and water loss from evaporation. An open pond loses about as much water due to evaporation as a grain field consumes in irrigation. Water loss increases retained salts and impacts culture stability. Some open ponds use seawater, waste or brine water, which often makes the water free but does not slow evaporation or salt concentration.

Algae Water remediation Ponds   *Plastic Bag*

Outdoor ponds make it difficult to control algae predators such as amoeba, ciliates, bacteria, rotifers, viruses, fungi, and zooplankton that can decimate the algae biomass within hours. [101] Open ponds are vulnerable to contamination from dust, windborne organisms, insects, and birds.

Commercial algae producers have devised strategies to minimize contamination by opportunistic weed algae in open ponds. Producers grow *spirulina* at high bicarbonate concentrations with high pH. *Dunaliella* grow in high saline water to discourage competing species. To manage contamination for *chlorella*, producers grow the biomass in batches with increasing volumes. Growers harvest the entire batch and the ponds purged and cleaned. Then new batches are restarted from clean laboratory cultures.

New research centers such as AzCATI at Arizona State University and algae incubators are testing various pond and photobioreactor systems to compare results and develop smarter automated systems. The pictures show two open raceways at the AzCATI test bed at Arizona State University.

Raceways at AzCATI built by Nano Voltaics

Most microalgae need light and carbon dioxide but they vary substantially by specie in nutrient and environmental requirements. Some species grow well in unlined ponds in Australia but the same variety may not flourish in unlined ponds in India or China. Local conditions often dictate the design and construction of open ponds as well as species selection and production methods.

The significant drawbacks of the open raceways have prompted the development of closed systems, called photobioreactors, made of transparent tubes or containers in which the culture is mixed by either a pump or CO_2 and air bubbling.

Closed microfarms

Covered, semi-closed or closed containers are designed to capture maximum solar energy. Systems may vary in size from a several square yards to several acres. Growing containers provide considerable visual variety and may be covered ponds, plastic bags, plastic sheets, flat plates, tubes or glass – anything that allows light to penetrate.[102] Some indoor systems use fiber optics or mirrors to capture sunlight or to add artificial light. Closed systems are more capital intensive than outdoor ponds.

Closed systems minimize contamination, permitting the cultivation of a single microalgae species. Closed systems offer better control over biocultural conditions such as pH, light intensity, carbon dioxide, nutrients and temperature. Tighter control lowers CO_2 losses due to out-gassing and minimizes evaporation.

Closed systems are often two to five times more productive per unit area than open microfarms. Closed systems offer the significant advantage of weather independence, allowing year round production. Growers counter cold weather and low light by adding sources of heat and light. Closed systems offer tighter control of contamination from unwanted algae, zooplankton predators, dust and debris. Closed systems may add cooling costs to guard against overheating in hot climates and higher cleaning costs from fouling.

Consumer research indicates creates negative consumer perceptions about the term bioreactor. Microfarm provides a constructive alternative. Photobioreactor implies that the sun excites plant cells to produce biomass through photosynthesis but naïve observer's associate reactors with nuclear power. Additionally, the term bioreactor has become synonymous with garbage waste disposal.[103] Consequently, the preferred terms are microfarm, biofactory or cultivated algaculture production system, CAPS.

Closed microfarm producers select algae species that maximize the characteristics desired such as biomass percentage of lipids, protein, or component product. Producers of food select to maximize biomass protein while nutraceutical growers select algae species with high lipid content. Growers have a new tool developed at Michigan State University to simulate growing conditions before investing in an algae growing system.

Phenometrics developed the computer controlled Environmental Photo BioReactor (ePBR™) for laboratories that simulates realistic environmental conditions such as temperature, light intensity and CO_2. The unit allows scientists to study algae under the same conditions found in various growing settings.

Each ePBR is a measurement instrument, quantifying growth rates, pH, and other factors in the algae culture, which are displayed on the computer monitor in a graph or data display. It is customizable to accommodate various probes and sensors that monitor algae culture and growth metrics. The ePBR™ provides reproducibility, a key factor in algae research experiments.

An array of Phenometrics ePBRs

In between laboratory simulators and field growing systems are laboratory systems that scale.

Professor Joel Cuello developed the Accordion PBR in the Biosystems Engineering Laboratory at the University of Arizona. The Accordion is a set of transparent bags that scale up to nearly any size to optimize culture productivity. The Accordion applies biochemical and ecological strategies for growing algae quickly, wastewater treatment and growing algae fertilizer for hydroponics.

The term "closed system" is a misnomer because algae predators and weed species invade any growing system. It may be better to think of a closed system as an arrangement that gives the grower more control over production parameters and more but not perfect control over contamination from invasive algae and predators. Growers typically control predators and weed species with a combination of parameters including pH, temperature, salinity and nutrients.

Tubular Microfarms

Ultraviolet (UV)-stabilized acrylic is typically used for construction because compared with glass; it is cheaper, stronger, lighter, more flexible, and easier to fabricate. Assembly of microfarms requires the integration of the various mixing, monitoring, and controlling subsystems.

Flat Plates built by Nano Voltaics for Arizona State University

Artificial lighting

Indoor growing systems with artificial lights allow growers to produce algae in any climate or geography 365 days a year. For example, Algaedyne's algae growing system allows farmers in Minnesota to grow algae on their farms.

At St. Cloud University, a student team analyzes the algae examines the nutrients recovered. The student team creates economic models for which recoverable algae components, protein, pigments, oils and other components offer the highest market value.

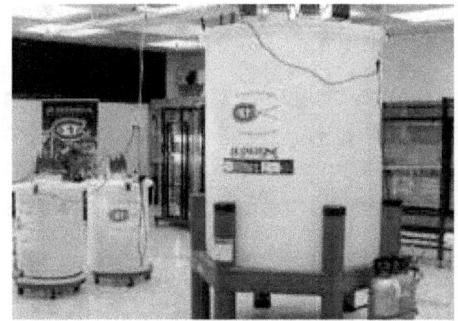

*Algaedyne Aglae System
at St. Cloud State University*

The interior light system produces more food on one acre than field crop farmers can grow on 80 acres. The system uses CO_2 given off by a farm's digester generator, which converts cow manure to methane gas. Manure nutrients are recycled to fertilize algae.

BioVantage Resources offers sustainable bioremediation solutions for industrial, agricultural and municipal wastewater treatment that may employ lighting. BioVantage recovers the nutrients from waste streams with biological solutions and turn the nutrients into high value products. The harvest biomass recovers resources such as phosphorus, metals, protein and energy, or for by-products such as biochemicals, bioplastics, biofuels and/or fertilizer.

BioVantage PBRs with Lighting   *Tank PBRs with Light Pipes*

The BioVantage integrated high efficiency LED lighting uses optimal and customizable wavelengths for photosynthesis. Light-pipe technology distributes illumination evenly throughout the water column for high-density growth, enabling more algae production within a smaller area. A patented pattern light-pipe illuminates the algae without adding heat so growers do not need heating or cooling.

BioVantage's bubble column and tank PBRs can be automated with medium preparation system and full growing system controls.

AlgaetoOmega developed unique technology for cultivating algae in controlled environments anywhere. Their business model is to convert vacant warehouses to scalable modular production systems that are low-cost, energy-efficient and environmentally sound. The system can grow any algae species in practically any location. AlgaetoOmega grows algae products for personal care, cosmetics, nutraceuticals and animal feed. The tubular PBR shown uses illumination from a LED panel behind the tubes. The solar tubes use solar energy delivered from the warehouse roof with fiber optics.

AlgaetoOmega Tubular PBR   *Solar Tube*

Vertical, lateral or angled tubular growing systems act as solar collectors. Some systems track the sun, similar to photovoltaic solar collectors. Horizontal closed systems, typically tubular or plastic bags, provide variations to solar exposure and production cost.

Tubular growing Systems

Different shapes provide distinctive benefits. A wide rectangle, similar to an aquarium, holds a lot of water but does not allow each alga cell to have sun exposure very often. Consequently, thin rectangular tanks, about four inches thick, tend to out-produce tanks that are wider. Tubular tanks may be a few inches wider because they present more surface area around the circumference. Tubular growing systems about six inches wide typically out produce wider tubes.

Experience with various microfarm designs in field settings shows the best architecture varies by setting and target production, Table 9.1. Algae cultures may need protection from ambient temperatures. Temperature spikes up or down may cause cultures to become unstable or slow growth.

Open pond	Economical, easy to manage, good for mass algae cultivation, considerable global experience and shared knowledge base.	Low culture control Stability issues Weak productivity High land and water use Species contamination Predator invasion
Vertical column	High mass transfer, good mixing with low shear stress, low energy consumption, scalable.	Small illumination surface Expensive construction Shear stress problems Cleaning issues.
Flat plate	Large illumination surface, good light path, good biomass productivity, relatively cheap, easy to clean, low oxygen build-up.	Scale-up challenges Culture stability Temperature stability Possible shear stress.
Tubular	Large illumination surface, good light path, relatively cheap.	Gradients of pH, dissolved oxygen and CO_2, tube fouling, high land use if laid flat.

Table 9.1. *Microfarm Types and Trade-offs*

Most algae cultivation systems can be covered. Covers give some protection against predators and weed algae species in open systems. Covers extend the growing season by retaining heat in cold settings and shading in hot conditions. Some growers flow water through an underground cistern to stabilize culture temperature.

The algae industry continues to experiment with variations in growing system design and operation. Low cost producers of commodities like feeds and fertilizers are likely to use open systems. Maximum total production may be achieved with hybrid systems where closed growing systems grow pure inoculate strains quickly to desired densities that flow to open or covered ponds for large-scale production and harvest.

Cost dominates the challenge of algae production. In order to be commercially viable, algae must produce biomass at lower dollar and energy costs than other food and energy alternatives. The National Renewable Energy Laboratory, (NREL), Algal species Program, for example, concluded in 1995 that closed systems were impractical for algae production because they were too expensive to build and maintain. Nearly all algae production to date occurs in open ponds. Growing strategies will change quickly since most planned algae production in the U.S. intends to use hybrid, closed or semi-closed production systems in order to improve quality control.[104]

Fermenters

Most algae species are autotrophs and use solar energy directly to produce organic substrates that store chemical energy from water and CO_2. Many species can also function as heterotrophs and are able to metabolize organic substances to create and store the chemical energy needed for their lifecycle. Heterotrophic algae can be grown in large containers called fermenters without light and are fed sugar as their primary energy source.

Fermenters offer several advantages including considerable published information on production as well as tested commercial growing systems. Heterotrophic growing systems may have lower operational costs than light-based systems, as long as a cheap sugar feedstock exists.[105] Algae biomass production without light usually uses a pure algae strain called an axenic culture that is free of other contaminating organisms. Molecular biologists have recently genetically transformed autotrophic algae that feed on light energy to heterotrophs. These transgenic cells thrive on sugar in the absence of light. The dual algae growth mode enables more flexibility in designing and operating growing systems.

Quality control

Measurements of process quality vary with the goal for the system. Optimizing food production requires substantial monitoring, testing and assurance that the process meets FDA and sometimes, organic food standards. Automated production systems enable quality control checks continuously for all the critical variables. Quality control may include monitoring for:

- **Biomass** density, color, size, structure, and vitality.
- **Water** temperature, pH (acidity), dissolved O_2, and CO_2.
- **Water** quality and dissolved salts and possibly metals,
- **Mixing** velocity and turbidity.
- **Nutrient** availability for all important nutrients,
- **Contamination** from weed algae or predator invasion.

Measurement of various component parameters needs to occur throughout the growing process as well as harvest, oil extraction, and component separation.

Production targets

Most of the planned production for algae biomass in 2013 targets large algae farms of several hundred acres to produce algae oils used as liquid transportation fuels. Algae farmed for energy promise extremely high return on investment, once technology challenges are solved. Liquid transportation fuels are the primary algae application receiving most public or private financing.

The algae industry will probably develop similar to traditional farming with small, medium, large, and mega-farms. Mega-farms will focus on liquid transportation fuels while other growers will cultivate a diverse array of algae products.

Superb graphics for real and imagined microfarms are shown in the highlights from the International Algae Competition, *Imagine our Algae Future,* (Henrikson and Edwards, 2012).

Imagine Our Algae Future

Chapter 9. Can Peace Microfarms Fight Hunger?

Hunger is actually the worst weapon of mass destruction. It claims millions of victims each year.
– Luiz Inacio Lula da Silva, President of Brazil

A strategy to prevent war with new forms of food from microcrops should fight hunger. America, the richest nation on the planet, provides an excellent case study. One in five children in the U.S. live in food insecure households where they get insufficient good food.[106] The USDA reports that over 17 million households, nearly one in seven, are food insecure in 2011. Food insecure household are up 30% since 2007, the highest number ever recorded in the U.S. Over 80 million Americans receive food support from the government or NGOs because they are hungry and cannot afford food.

Unfortunately, these sad statistics will escalate due to the widespread agricultural losses from a summer of severe drought and heat. Over half the counties in the U.S. were declared disaster areas by President Barack Obama due to soaring temperatures that destroyed crops.[107] Many of the rural poor depend on their small farms for food and income to buy food. When weather ruins their crops, they have nothing to eat and nothing to sell.

Hunger in America offers a microcase for world hunger. Over 2 billion people globally suffer from hunger and malnutrition. The same Green Algae Strategy proposed here holds promise for world hunger.

The cost of hunger

In *Ending Hunger Now,* former Senator, Presidential candidate and United Nations ambassador on hunger, George McGovern says that the cost of hunger is unacceptable.

> *Today's malnourished pregnant and nursing mothers are producing tomorrow's barriers to personal, social and economic development – malnourished, brain dulled, listless children. Those fortunate enough to survive will go through an uncertain life, permanently diminished and unable to be productive, happy human beings.*[108]

Senator McGovern is absolutely right about the unacceptable cost of poverty. America's hunger bill costs our nation over $168 billion a year from the combination of lost economic productivity, drag on public education, avoidable health care and the cost of charity to keep families fed.[109]

Added to this $168 billion are the $94 billion a year in the USDA Supplemental Nutrition Assistance Programs and the other key federal food assistance programs:

- Supplemental Nutrition Assistance Program, (new name for food stamps), $78 billion.
- Special Supplemental Nutrition Program for Women, Infants and Children, $7 billion.
- National School Lunch Program, $14 billion.[110]

Even if spending over $250 billion a year to support America's poor and hungry made economic sense, it undermines family life and deprives many of our children the opportunity to participate in the American Dream. Many of our 20% food insecure children will not advance in school, will create huge healthcare costs and will not be strong enough for military service.

Federal food assistance programs began during World War II when 40% of American young people suffered from nutritional deficiencies and could not qualify for military service.

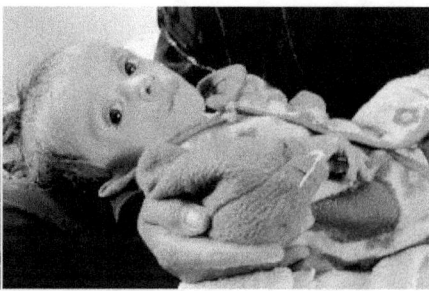

Faces of Hunger

Ending childhood hunger should be America's number one strategic priority. Unlike climate change, there is no ambiguity that childhood hunger is human caused and is increasing with catastrophic consequences.

Health consequences of hunger

Nicholas Kristof wrote in the New York Times:

> The most heartbreaking thing about starving children is their equanimity. They don't cry. They don't smile. They don't move. They don't show a flicker of fear, pain or interest. Tiny, wizened zombies, they shut down all nonessential operations to employ every last calorie to stay alive.[111]

Many children suffer from hidden hunger because they do not get sufficient micronutrients — iron, zinc, vitamin A, iodine. Micronutrients give produce their color, taste, texture and aroma. Micronutrient deficiencies zap the color and vitality from children's lives.

A study released by the Center on Hunger and Poverty at Brandeis University and the Food Research and Action Center (FRAC) shows hungry children compared with children with enough food, suffer from two to four times as many health problems.[112] Maladies included unwanted weight loss, fatigue, headaches, irritability, inability to concentrate and frequent colds. The relationship between hunger and health problems was unaffected by income. Hunger has a strong effect on children's health no matter what the income level of their families.

According to the FRAC report, hungry children are more likely to be ill and absent from school. Imagine the drag on education when classmates are hungry, obese, irritable and unable to focus on assignments. Constant fatigue ruins family life and diminishes the child's ability to learn.

Inadequate nutrition causes stunting in children. The Surgeon General's 1990 goal of eliminating growth retardation of infants and children caused by inadequate diets was not met because significant numbers of low-income children continued to suffer retarded growth.[113] Slow growth also affects the vital organs including the brain, eyes, heart and circulatory system.

Iron-deficiency anemia occurs from insufficient iron. The body cannot produce enough hemoglobin, the substance in red blood cells that enables them to carry oxygen. As a result, iron deficiency anemia leaves a person tired and short of breath.[114] Iron-deficiency anemia leads to developmental and behavioral disturbances that diminish children's ability to learn to read or do mathematics. Anemia remains a significant health problem among low-income children and women. The Centers for Disease Control reports that in 2012, about 14% of infants 1 to 2 years old suffer from anemia. Over 9% of women 12 to 49 suffer from anemia, which leads to more illness and additional and longer hospital stays.[115]

Vitamin A deficiency leads to abnormal bone development, disorders of the reproductive system and dry eyes. Lack of Vitamin A causes loss of color vision, then night vision, and finally blindness.[116] Blindness is typically followed by death. People get Vitamin A from leafy vegetables, fruits and carrots. Carotenes are provitamins because they can be converted to active vitamin A. Carotenes possess antioxidant properties, which protect against the effects of free radicals. Free radicals are molecules produced when your body breaks down food, or by environmental exposures like tobacco smoke and radiation.[117] Free radicals damage cells, and may play a role in major organ diseases including heart, cancer cardiovascular, Parkinson's and Alzheimer's.

According to the Tufts University Center on Hunger, Poverty and Nutrition Policy, inadequate child nutrition imposes detrimental effects on the cognitive development of children and results in lost knowledge, brainpower and productivity for the nation.[118]

Hunger and malnutrition exacerbate chronic and acute diseases and speed the onset of degenerative diseases among the elderly. This not only leads to an unnecessary decrease in the quality of life for many older people, but also increases the cost of health care.[119] National data for people over 65 show that a majority are not consuming even two-thirds of the nutrients they need to stay healthy.

Obstacles to good nutrition

Why are children and the elderly not getting good nutrition? The root causes are access, hidden hunger, (nutrient deficiencies) and food prices. Many people live in food deserts where good food is simply not available locally.[120] Food deserts are areas that lack access to affordable fruits, vegetables, whole grains, lowfat milk, and other foods that make up the full range of a healthy diet. The USDA provides a handy food desert locator.

Many industrial foods suffer from hidden hunger, which results in foods with empty calories.[121] Hidden hunger refers crops that need more of one or more nutrients, yet shows no visible deficiency symptoms. The nutrient content is above the deficiency symptom zone needed for passable appearance but below the zone for optimal crop health.

Crops give farmers a relatively wide zone where withheld nutrients, (fertilizers) show no visual effects. Farmers know that extra nitrogen creates larger produce even when several micronutrients are low. Consumers are attracted to larger produce and few suspect that the extra weight comes largely from water, not nutrient-rich biomass.

Farmers are paid by yield, not nutritional availability or density. Consequently, many industrial foods are nutrient deficient because farmers hold back fertilizers save to save money. For many farmers, fertilizers represent 40% of the cost of their crop. The result is empty calories, produce that delivers few nutrients per bite. Processing food with extra sugar and fat make food attractive to children but amplifies empty calories.

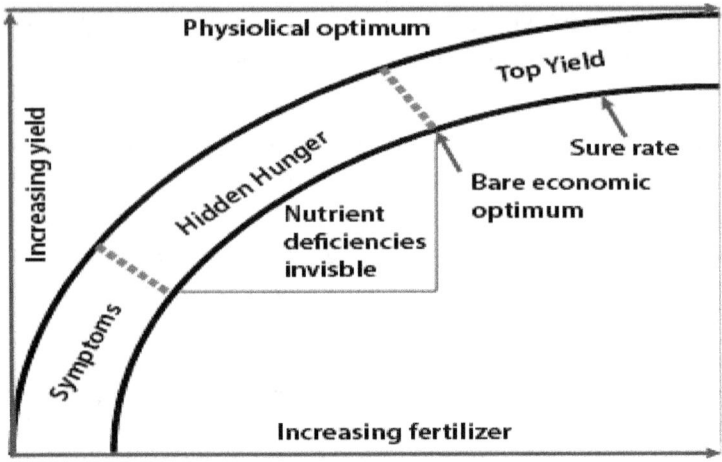

Hidden Hunger in Industrial Foods

Hidden hunger in foods transfers hidden hunger to people. While produce with hidden hunger may not show visible effects, hidden hunger in people often results in both visible, (stunting) and behavioral problems (inability to concentrate) associated with nutrient deficiencies.

Food costs prohibit many from obtaining good nutrition.[122] Food costs rise with the cost or fossil resources, especially fuel and fertilizers. Food prices increase when climate change causes diminished yields or crop failures. Transportation costs also escalate the cost of food because the average American foodstuff travels an estimated 1,500 miles before being consumed.[123]

Green solution

People need access to lower cost fresh local foods that are packed with nutrients. Peace microfarms use abundance methods that assure sustainable food production for many generations. Growers produce freedom foods that give consumers choice for healthier foods.

Access	Peace microfarms enable microfarmers to grow superior foods close to consumers in cities, towns, slums, barrios as well as rural areas. Freedom foods are climate independent, which means growers can provide fresh foods that are local to consumers, all year round. Many families can grow their own food in a microfarm or grow and flow algae feed to their animals or rich algae biofertilizer to their gardens.
Hidden hunger, nutrient deficiencies	Growers produce freedom foods that give consumers choice for healthier foods that fight diseases. Freedom foods offer 200 to 500% higher nutralence – nutrient availability and density – than industrial foods. Carrots are the highest source of beta-carotene among terrestrial foods crops. Each gram of algae contains 10 times more beta-carotene than a gram of carrot. One tablespoon of algae a day stops hidden hunger and nutrient deficiencies such as iron, zinc, vitamin A and iodine in children and adults. Algae biofertilizers can also end hidden hunger in field crops.[124]
Food cost	Peace microfarms use abundance methods, which frees growers from the continually rising cost of fuels and agricultural chemicals. The recovery and recycling of nutrients from waste streams saves growers much of the fertilizer cost. Green energy saves fuel and electricity. Abundance methods save on pesticides and agricultural chemicals. Algae offer natural biodiversity, which saves growers from the rising cost of genetically engineered seeds. Local food production lowers the cost of food substantially by reducing or eliminating processing, preserving, packaging, storage, transportation and handling,

A Green Algae Strategy to Resolve Hunger and Malnutrition

Peace Microfarms

Microfarms can make fresh healthy foods available; give relief from nutrient deficiencies and lower food cost. All that remains is our willingness to take action.

Chapter 10. Why not Now?

Modern foods seem cheap because neither farmers nor consumers pay for extraction, consumption or pollution.

T

The first question most people ask about peace microfarms is, "Why weren't microfarms using abundance methods invented before?" The answer: poor math. Industrial foods appear to be cheaper because they do not reflect the true costs. The cost of subsidies, resource depletion, fossil energy consumption, waste, pollution and health care are ignored, and transferred to our children. Consumers will benefit from full natural resource and lifecycle accounting.

Intellectual property, (IP) has hindered new foods too. Peace microfarms mimic the "natural process," developed by nature, not by man. Natural processes belong to everyone and cannot be patented or listed as IP. Agribusinesses that supply inputs to farmers hire legions of technologists and scientists to develop synthetic compounds. These patented compounds, synthetic poisons, and GE seeds sell to farmers for huge premiums that create wealth for the companies and their executives. Agribusiness advertising promotes this intellectual property and creates a belief in better living and farming through chemistry. Monsanto's Round-Up™ is among the most common words in modern farming.

Consumers are addicted to highly processed foods, with high sugar, fat, cholesterol and salt but few nutrients. Farmers are addicted to increasingly expensive GE seeds, synthetic fertilizers, and patented herbicides, pesticides and fungicides. The sad irony is that the synthetic poisons kill the beneficial microbes that nature put in the field to feed and nurture plants. Imagine, paying premiums to kill the organisms that work symbiotically with plants to provide nutrients and plant hormones for their vitality and defense.

Modern farmers have bought into chemical fertilizers because they are easy to apply and decades ago they were cheap. Industrial agriculture systemically extracts macro and micronutrients as well as organics from field soils. Each year, many farmers replace primarily the three N-P-K macronutrients (nitrogen, phosphorus and potassium). With every crop, micronutrients diminish along with soil organics, which creates nutrient dilution and hidden hunger.

Modern farmers use large, heavy tractors that cultivate quickly but compact soil, which diminishes root growth and accelerates erosion. Farmers buy tons of chemical fertilizers, pesticides, herbicides and fungicides. Crops are developing resistance to chemical fertilizers, so farmers must apply more. Pests and weeds are developing resistance to chemical poisons, which means farmers must use more or change poisons.[125] Plants often absorb less than 5% of the agricultural poisons applied to fields, which creates enormous waste and cost. The residual fertilizers and poisons flow into wetlands, streams and groundwater where they damage and destroy local ecology.

Fertile soil is not an inert medium but a mixture of water, air, minerals and organic matter. In most soils, minerals represent around 45% of the total volume, water and air about 25% each, and organic matter 2-5%.[126] The mineral portion consists of three distinct particle sizes classified as sand, silt or clay.

Soil health depends on the organic component that house many living creatures along with dead material in various stages of decomposition. An acre of living soil may contain 900 pounds of earthworms, 2400 pounds of fungi, 1500 pounds of bacteria, 133 pounds of protozoa, 890 pounds of arthropods and algae, and possibly some small mammals.[127] An acre of soil may contain over 10,000 species of microorganisms, which contributes to the biodiversity in living soil.[128] Unfortunately, industrial agriculture acts to kill the microorganisms with cultivation, soil compaction, chemical fertilizers and agricultural poisons.

Soil organic matter is the smallest but most critical soil component for crops. Soil organic matter interacts to influence soil biological, chemical and physical properties and consists of raw plant residues and microorganisms, (1-10%); active organic traction, (10-40%); and resistant or stable organic matter, (40-60%) called humus.[129] Modern farmers replace the macrofertilizer, but not the humus removed by each crop.

Peace microfarms are antithetical to agribusiness firms because foods grown with "natural processes" are not patentable. Similarly, bioavailable algae fertilizers and other plant inputs are natural and are not patentable. Nature engineered marvelous products that provided for plant needs eons before Monsanto entered the business. Algae and the symbiotic microbes they attract create provide the compounds that enable plants to naturally synthesize many of the advanced compounds they need to grow and to fight disease and pest vectors. Unfortunately, U.S. government farm policy chose to support R&D on industrial food production rather than natural processes.

Farm policy

Food production is dictated by farm policies. Government sponsored research to Land Grant Universities, extension service agents, subsidies, and food support for the hungry are governed by farm policy. The same large agribusinesses that have addicted farmers to their branded synthetic chemistry drive farm policy. Wealthy farmers and agribusiness like ADM, Monsanto and Cargill make enormous political donations to both parties in order to shape policy in commercial agriculture to benefit their interests. Consequently, over 99% of federal grants and R&D go to industrial agriculture. Unsurprisingly, most extension agents who are in place to help farmers and gardeners receive training in industrial production. Less than 1% of U.S. federal funding goes to organic production.

Government funding for natural growing methods such as abundance and freedom foods rounds to zero.

India and China support natural processes R&D in food production because their leaders realize that fossil resources are finite, increasing in price, and will eventually run out. Both countries have terminated their biofuels programs with food crops for the obvious reason that food-based biofuels drive up the cost of food and the inputs to produce food.[130] China recently put a 135% tariff on their phosphorus fertilizer to insure sufficient supplies for domestic farmers. India's scientists have performed some excellent R&D with natural biofertilizers, especially focused on cyanobacteria that fix nitrogen and reduce the need for nitrogen fertilizer.

As modern farm policies evolved from 1960 to 1990, food supply and sustainability issues were not well articulated. Consumers and political leaders preferred celebrating their brilliance in designing the Green Revolution and the cheap food it provided. Leaders and policy makers ignored critical issues with GE crops, such as the need for additional cultivation, two to three times more irrigation, triple the need for fertilizer and ten times the need for agricultural chemicals.

Few people were aware of nutrition and health issues, food security, fossil resource depletion or global warming before the 1980s. The winds of political rhetoric drowned out the few voices that challenged the fossil foods path such as Prince Charles, Vandana Shiva, Michael Pollan, Miguel Altieri, Alice Waters, and Robert Henrikson.

Today, only one third of Americans believe the scientific consensus that human actions cause global climate change. However, neither politicians nor consumers can deny that humans have caused severe fossil resource depletion and environmental pollutions with our cheap fossil food policies.

Cheap food?

Freedom foods make little sense when the cost consumers pay for industrial foods appear to be so cheap. Appearances can be deceiving.

The U.S. government lavishly subsidizes industrial farming, big agribusinesses, big oil, water management and the fossil resources on which food production depends. For example, many farmers pay less than 2% of the true cost of irrigation water – which promotes waste and pollution. Subsidies reduce the real food cost by nearly a third, Figure 10.1. Subsidies are financed with our children's money, in U.S. bonds held by countries like Saudi Arabia, Egypt, Qatar and China.

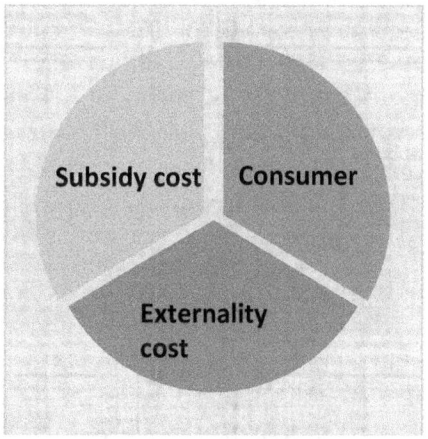

Figure 10.1 Real Cost of Food

American corn subsidies decimated Haitian farmers because they could not grow food as cheap U.S. food dumped on the country as "food aid." The U.S. corn subsidies also have displaced over a 1.5 million poor Mexican farmers. Farmers were forced to leave their land because they could not compete with subsidized U.S. corn. Many of these farmers added their feet to the flow of illegal immigrants to the U.S. from Mexico. Canada, Mexico and other countries have outstanding lawsuits against U.S. subsidies with the World Trade Organization because these subsidies substantially depress the real price of food grains.

A group of more than 400 agricultural experts, known as the International Assessment of Agricultural Knowledge, Science and Technology for Development concluded through its global and regional studies report that governments and industries need to discontinue environmentally damaging farming methods. At their 2008 meeting in Johannesburg South Africa, the group recommended, "ending subsidies that encourage unsustainable practices." Political leaders in the U.S. should listen to world opinion because U.S. subsidies amplify resource consumption and pollution. Subsidies today will destroy our ability to grow our own food in the near future.

Another third of the true food cost comes from externalities such as resource depletion, environmental degradation and human health impacts, for which the food supply system fails to account. Environmental degradation alone creates about $45 billion a year in damage. No metrics are currently available for resource depletion. Neither farmers nor consumers pay a nickel for these costs. These hidden costs are shifted to our children. Unfortunately, when the groundwater crashes in the Midwest, our children will not be able to buy sufficient water at any price.

A full lifecycle accounting would show fossil foods are substantially more expensive than freedom food production. Life cycle accounting includes the economic impact of degrading air, water and soil, destroying our fisheries, creating dead zones as well as cost to human and animal quality of life and health. The current generation benefit from over-consuming fossil resources and polluting ecosystems. We ignore resource loss by failing to account for depletion in the price of our food. The next generation will not enjoy the same luxury.

Prince Charles in his Future of Food speech at Georgetown University pointed out the "curiously perverse" economic incentive system (subsidies) that too frequently directs food production. He addressed the true cost of food effectively:

> *Nobody wants food prices to go up, but if it is the case that the present low price of intensively produced food in developed countries is actually an illusion, only made possible by transferring the cost of cleaning up pollution or dealing with human health problems onto other agencies, then could correcting these anomalies result in a more beneficial arena where nobody is actually worse off in net terms? It would simply be a more honest form of accounting that may make it more desirable for producers to operate more sustainably, particularly if subsidies were redirected to benefit sustainable systems of production.*[131]

Prince Charles recommends "accounting for sustainability," which represents the true cost of food production, financial costs and the costs to natural capital – the earth's resources.

When our children discover industrial agriculture lacks the natural resources to produce food, they will ask the government for increased subsidies. Unfortunately, the government will be out of funds. What country would be willing to make loans that add to the immense U.S. debt? The U.S. is already a debtor nation; we just fail to act as one.

When our children discover their fields worn out, fresh water is unavailable, fuel costs are out of reach, fertilizer mines are exhausted and agricultural chemicals have ruined their waterways – will they agree that our fossil foods were cheap?

Biofuels

In the 1990s, the Clinton administration made a critical political mistake and stopped R&D on algae for food or biofuels. Those funds were shifted to corn ethanol for biofuel. The decision by the EPA to fund a corn ethanol industry may have been the most costly decision in American history because it accelerates natural resource depletion. When the U.S. runs out of resources to produce food, who will sell us food? Where will the government find the money to buy food for hungry Americans?

The farm lobby remains so strong that corn ethanol subsidies continue at around $6 billion a year, even though ethanol consumes more fossil energy than it returns. Huge subsidies flow primarily to large agribusiness and landowners, not to family farms. Subsidies continue in spite of clear scientific proof that corn ethanol is an expensive, wasteful proposition that not only massively depletes our natural resources but creates billions of dollars in degraded and damaged ecosystems. The 44 million acres of corn grown for ethanol in 2010 could and should be replaced by less than 2 million acres of algae biofuel production, while improving air and water quality.

Biowar I: Why Battles over Food and Fuel Lead to World Hunger (Edwards, 2007) traces the money path, primarily to one company, ADM, that initiated the biofuel industry with millions in political donations to both parties. Today ADM receives billions each year in biofuel subsidies. A biowar occurs when a country burns food, typically as a horrific act of war on another country.

In Biowar I, the U.S. became the first country to burn its own food. Biowar I ignited when the Bush Administration announced the Energy Policy Initiative in 2005, which increased biofuel subsidies and mandates, Figure 10.2. The unintended consequence of producing large amounts of corn ethanol on U.S. and world food markets was predictably higher food prices. In the eyes of the UN, World Bank and most foreign countries, the U.S. ethanol policy contributed substantially to the terrible 2008 food riots in 40 countries.

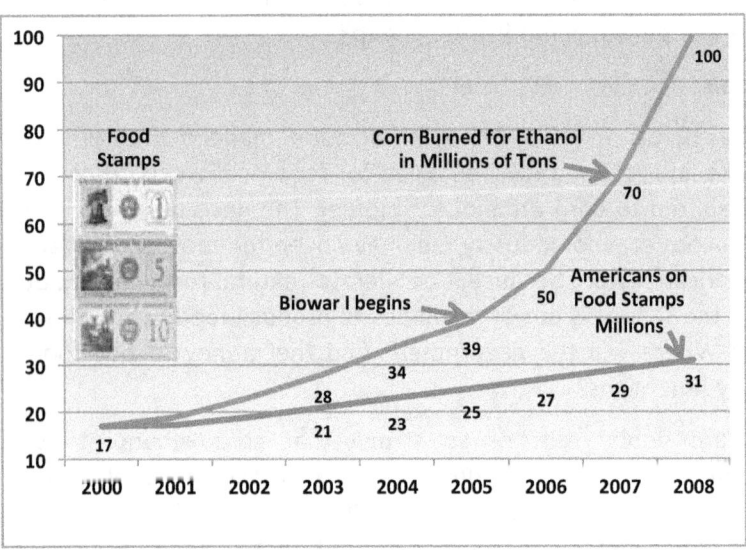

Figure 10.2 Corn Burned for Ethanol and Food Stamps

How could people in hungry countries not blame the U.S. for food shortages and price increases when prior to the ethanol program, America provided half the world's food grains and 70% of the world's corn imports? How does a country with over 60 million people receiving food support because they are hungry justify a policy of burning its citizen's food for a weak fuel additive? Over 48 million Americans are on food stamps and must abhor the concept of burning food because they know their $1 a meal buys less food.

In 2009, the U.S. became a net importer of food. A college sophomore could make the case that the U.S. biofuel policy is wasteful and foolish. We are burning our children's natural resources.

We may have developed a less expensive, food production system had people not had distaste to one word – algae.

Consumer behavior and algae

Why do most people have an immediate aversion to algae? The answer is false attribution. When asked to describe algae, people's top of mind typically elicits several words with strong negative connotations: "slimy, smelly, scummy and yucky." If putrefied raw meat were presented as steak, people would naturally dislike steak.

People falsely attribute the smell in ponds to algae because it certainly looks like algae. Actually, the odor comes not from algae but from the bacteria that attack and eat the algae. The bacteria consume all the oxygen algae added to the water, causing eutrophication. When the aquatic organisms are deprived of oxygen, they die and begin rotting, which adds smell and slime to ponds. Healthy algae give off lots of oxygen and smell similar to walking through a redwood forest – without the redwood trees, of course.

Weed algae in ponds grow in diverse communities of many microorganisms and are different from the algae we cultivate in abundance microfarms. Gardeners know they must remove the weeds from the garden or the weeds will take over the garden and consume all the nutrients. Abundance growers control weed algae in order to enable healthy production of the target species.

Consumers enjoy the taste of edible algae. At a recent Rotary meeting at a Kobe Steakhouse, the Rotarians were served three forms of algae at the luncheon: algae soup broth for taste and thickening and sea vegetables in the soup for texture, color and visual appeal. Separately a cold seaweed salad was served for color, taste and texture. The Rotarians were asked after lunch if anyone liked algae to eat, which elicited the yuck factor, as all were extremely negative. They reversed judgment when they understood sea vegetables and algae salad provided the colors, taste and texture to their lunch. The Kobe manager brought out a large variety plate of sushi made with algae. The Rotarians consumed the sushi plate in one pass.

Nori, with a world market value of over $3 billion a year, represents only one of more than 1,000 sea vegetable. Nori serves as a luxury food. Cooks wrap Nori around a rice ball with seafood on the top. Toasting or baking brings out Nori's rich flavor and flakes complement rice or noodles. Epicurean cooks make a Nori soy sauce reduction for meat or seafood. Nori adds a spicy taste to jam and wine. Many other sea and freshwater algae foods await commercialization.

Many forms of algae do not taste good by themselves. Some algae may need processing with other ingredients. Raw soybeans taste terrible but they are palate pleasing when processed to tofu. Algae-based foods will become more attractive with informative food labels.

Food labels

Current food labels display only a few bits of information about calories and fat. We propose a more comprehensive food label that represents the nutritional value, sustainability, social value and ecological costs of food. Since the label is too long for many food products, we are working on a smart phone app that will allow consumers to access extensive label information for any food product.

We plan to expand the concept of taste tests to demonstrate the benefits of freedom foods. These blind taste tests will ask consumers first for their preferences based on standard consumer behavior protocols. A secondary test will ask for consumer preferences for food A or B using snack chip label similar to Table 10.1.

Expanded food labels will provide consumers with additional information that is currently unavailable. Consumers currently have no information on food labels that indicate food subsidies, fossil resource consumption, GE or GMO, food miles, waste or pollution. Some consumers will ignore the expanded labels, just as they ignore the current brief labels. The expanded label strategy will encourage dialogue and debate on sustainable and affordable food and energy (SAFE) production.

Food type	Industrial food	Freedom food
Nutrients	Corn chip	Algae chip
Protein	5 g	41.4 g
Calories	432 g	188 g
Calories from fat	16 g	3.7 g
Saturated fat	5.4 g	0.55 g
Polyunsaturated fat	3.9g	1.39 g
Monounsaturated fat	2.5g	0.20 g
List other micronutrients		
Health and nutrition		
High nutralence	No	Yes
Health risk from GE	Unknown	No
High in vitamins and minerals	No	Yes
High in antioxidants	No	Yes
High in trace elements	No	Yes
No pesticide residue	No	Yes
Processing and preservatives	High	Low
Environmental health		
GE crops	Yes	No
Monoculture	Yes	No
Fossil resource consumption	High	Low
Food transportation miles	High	Low
Waste of natural resources	High	Low
Ecological pollution	High	Low

Table 10.1 Expanded Food Label
Example: 100 grams of Corn and Nori Chips

Freedom foods are not a panacea and offer solutions to only some food issues. A set of needed technology breakthroughs, defined in the DOE National Algal Biofuels Technology Roadmap, will be required for the optimum use of algae biomass for commercial production.[132]

Chapter 11. Peace Microfarms FAQ

Peace microfarms will have some unintended consequences.
We must actively monitor progress and insure transparency.

If designing, developing, demonstrating and diffusing peace microfarms were easy, they would have been built decades ago. We are just beginning the development of an exciting new growing method. Abundant agriculture 1.0 requires considerable additional development. Initial questions include the following.

1. **How could peace microfarms save 100 million people by 2030?**

   Peace microfarms can save millions of people from starvation, malnutrition and war by:

   a. Enabling people globally to create a healthy food supply grown locally with plentiful rather than fossil resources.
   b. Preserving fossil resources so there is no need to fight over them, especially cropland, water, fuels and fertilizers.
   c. Cleaning waste streams and polluted ecosystems while recovering and recycling valuable nutrients.

2. **Farmers will never give up dirt farming and adopt microfarms to grow freedom foods.**

   Farmers do not have to abandon their fields or their favorite crops. Farmers can regenerate their fields and improve crop yields. Growers can leave their fields better than they found them.

3. We will not run out of fossil resources.

Really? The question is not whether we will run out, only when. In many developing countries, critical resources, especially productive seeds, fresh water, fuel and fertilizer have become too expensive. Read Michael Klare, *The Race for what's Left: The Global scramble for the World's Last Resources*, 2012.

About 27% of U.S. irrigated farmland depends on the Ogallala aquifer, which is dropping up to two feet per year.[133] Parts of the Ogallala will go dry within 30 years and it would take about 6,000 years to recharge with rainfall.[134]

4. How will we pay for transforming our food system?

Modern industrial foods have benefited from tens of billions of dollars in R3D, education and subsidies. Microcrops have received practically no government support and no education or subsidies. Algae production for biofuels has received less than $1 billion in government R&D funding but has not addressed food production.

The freedom foods industry can begin as a supplemental food supply that does not compete with industrial foods. Policy leaders may shift subsidies from fossil foods to ecologically positive foods. Local policy leaders may begin taxing or penalizing polluters who degrade our ecosystems for this and the next generation. Policy leaders may decide to shift budgets from less strategic projects to sustainable food production and food security.

5. Algae production is still 10 years away.

No, firms have been producing algae profitably for 30 years. We conducted the only algae industry surveys to date, (for the Algal Biomass Organization and for Algae World in Asia). Those results show that over half of producers believe successful algae production is operational now. Respondent see algae production distributed in nearly all geographies. Over 150 companies are producing algae successfully today. Once peace microfarms are available for gardeners, growth will become exponential.

6. Algae production is too hard.

Yes, microfarms have been too difficult, but that is changing. Both authors have failed multiple times over the last 30 years constructing backyard systems. All we need is one success – and a person or team willing to share the techniques as open source. We are engaging Green Masterminds, engineers, architects, and scientists at www.AlgaeCompetition.com to share their innovative designs and production breakthroughs. Please join the fun and become a full-fledged Green Mastermind.

We have successfully prototyped smartcultures production near fields. Crop yields and cost reductions exceeded our expectations. However, we need to make the systems less expensive, more portable and easier to operate and maintain. These upgrades will happen soon. Diffusion before microfarms operate reliably and efficiently would create the same false promises that doomed many prior global food initiatives.

7. Algae production is too expensive.

No, fossil agriculture is too expensive because consumers pay only one-third the cost of food. One-third is masked by subsidies and another third, ecological cost, is unaccounted for and ignored. A lifecycle analysis of fossil foods that accounts for all costs including extraction, depletion, waste, erosion, ecological pollution and all the impacts to human and animal health, will show that industrial agriculture is more expensive than freedom foods. The key political question will be how long people are willing to ignore the substantial subsidy and ecological costs imposed by industrial agricultural production.

8. Aren't microfarms simply organic food production?

Microfarmers may use organic or industrial methods to cultivate foods low on the food chain, microcrops. Organic growers of land-based crops avoid, to the degree possible, fossil inputs such as GE seeds, chemical fertilizers, and agricultural poisons. However, organic production typically requires more fertile soil, freshwater and fossil fuel than industrial agriculture. Both organic and industrial growers can adopt abundance methods to improve yields, reduce costs, and operate more sustainability.

9. If freedom foods were this good, they would already available.

No. Prior to recent biotechnology breakthroughs, freedom foods were impractical. Recent innovations in biophysics, biochemistry, bioengineering and a host of other disciplines have converged to make abundance methods possible now. Of course, we will have to apply new technologies to make production easy and reliable.

10. There is no such thing as a free lunch.

This critique represents the food chain with the little fish eaten by successively larger fish. Freedom food production is not free in an economic sense because all inputs have some cost in terms of transportation, labor or capital. Recovering and reusing farm waste stream nutrients closes the nutrient cycle and frees farmers from continual extraction from their fields. It frees farmers from the need to continually purchase more fossil fertilizers and chemicals. While abundance may not be free, the practice offers a very cheap lunch compared with the escalating costs of modern fossil foods.

11. People may not want to eat algae.

True, some people will continue to eat meat and potatoes, assuming they can still afford the price of meat. Some may have cultural values that forbid ingesting microorganisms – other than the algae, fungi, bacteria, molds, mildews, yeasts and viruses that ride naturally on everything we eat. These organisms also reside in our gut, where they work symbiotically to provide bioavailable nutrients from the food we ingest.

People who want to abstain from eating algae directly may use abundance methods to grow algae for aquaculture, hydroponics or aquaponics. Others may grow algae to feed their birds, dairy or meat animals. Farmers may grow algae in the smartcultures model to improve yields and quality of their field crops.

New food processing technologies will transform empty calorie snack foods into tasty, low-fat, high nutrient health foods. A food renaissance will transform convenience foods into foods that build strong bodies and minds. Algae flour and oils will enable people to have their chocolate cake with ice cream and whipped cream – and eat it too – without fat and calorie guilt. Superb algae-based gourmet foods will become the rage in upscale restaurants.

12. What if algae run amok?

Algae are already plentiful in the natural environment and produce 40% of the world's new biomass daily. Should some algae get loose, the culture will grow to the limits of the in situ nutrients and stop. Growers can kill algae with chlorine or copper, as they do in pools and aquariums. A Google.com search on algae provides 10:1 information on how to kill algae rather than how to cultivate algae.

13. What if algae carry a disease or create toxins?

There is a legitimate concern that algae, just like any other food, may carry salmonella, e-coli or another pathogen. Continual monitoring, possibly remotely, can identify and stop cultures that may be problematic. The research on algae toxins identifies the species and growing conditions necessary to produce toxins. Growers avoid both the species and growing conditions. Intense R&D continues on algae toxins will continue because algae toxins have stopped tumor cell reproduction for 30 types of cancer.

14. Aren't waste streams dirty?

Yes, and the pathogens can be killed using natural tools such as UV light and solar heaters. These technologies have been used successfully for 50 years throughout the food supply chain.

15. Argh!

Algae are just a bunch of cells. Yes, just like a carrot but with 10 times more beta-carotene per ounce. Carrot cells differentiate themselves into root, stems and leaves, while alga cells do not. Macroalgae are made of many independent undifferentiated cells often organized as pseudo-stems and leaves, similar to land plants.

16. Peace microfarms do not address population management.

True, and without some form of population management, many are doomed to poverty and hunger. Microfarms address large families indirectly because families in developing countries often have extra children in order to assure labor for the farm. Microfarms remove the need for extra children as a form of supplying farm labor because abundance enables food production with relatively light work that can be performed by almost anyone.

17. Will microfarms foods take jobs away from farmers?

No, abundance methods engage farmers in a new set of actions that makes farming more profitable. Many aspiring farmers can practice farming because microfarms require only modest space. Microfarms will create a new industry with millions of green jobs.

18. How can we get started now?

Use social media to explore your ideas. Create teams that design environmental landscapes, microfarms, algae foods and menus. Post your ideas at www.AlgaeCompetition.com. Use the site to extend and refine your concepts and find other like-minded social entrepreneurs.

19. The farm lobby is too strong with its huge political lobbies and guaranteed subsidies.

The farm lobby shames the U.S. the Congress and all Americans. Subsidies pay enormous amounts of public money, 85% of which goes to large agribusinesses and a few wealthy landowners – not hard working family farmers. The farm lobby may embrace freedom foods in order to answer to Congress for the health of our children, the overconsumption and loss of our precious natural resources, and the pollution of our ecosystems.

20. No highly productive, low-cost, easy maintenance microfarms are on the market today.

True, which means we must design, develop, demonstrate and diffuse microfarms that anyone can use.

21. There are no businesses to support freedom food production.

True today, but consider the many phenomenal entrepreneurial opportunities in this new industry. The industry needs designers, builders and input suppliers. Growers will need trainers, growing support systems and harvesting technologies.

22. Could subsidies to farmers to help a transition to abundance?

America needs is an agriculture or homeland security secretary who has courage to make a smart policy statement:

> *Food and energy subsidies should benefit the production of sustainable and ecologically positive farming practices.*

Congress could then shift the $6 billion a year in subsidies and costs currently wasted on corn ethanol to multiple forms of truly sustainable energy.

23. You do not address vegetables versus meat.

True, we elected to leave those arguments for others. Others have written terrific books and articles on the value proposition for eating lower on the fossil food chain with vegetables. Notable works include the *Moosewood Restaurant Cooks at Home: Fast and Easy Recipes for Any Day* by the Moosewood Collective, *Quick-Fix Vegetarian: Healthy Home-Cooked Meals in 30 Minutes* by Robin Robertson, and *How to Cook Everything Vegetarian: Simple Meatless Recipes* by Mark Bittman and Alan Witschonke.

24. Some people will always prefer meat.

Meat eaters will soon have a choice of a $35 beef hamburger, a $15 tissue-culture (pink slime) burger or an $8 algaeburger made with texturized vegetable protein. Many meat eaters will choose the algaeburger because it delivers better taste as well as healthier fat and protein with only a fraction of the cholesterol from beef. The Umami Burger chain is among the fastest growing food retailers. Resource costs will drive up the cost of meat animals.

25. You will never be able to train people to grow algae.

We do not intend to train just anybody. We plan to train Green Masterminds, who share freedom food passion and are willing to conduct peer training with others. We need first to train farmers and master and hobby gardeners with deep knowledge about growing plants. We will need freedom foods demonstration sites in every community, so there will be many grower opportunities.

26. FDA and USDA policies and certifications will inhibit adoption.

True. Regulatory procedures for food and feeds are onerous and expensive. The FDA requires GRAS (Generally Regarded As Safe) certification for algae foods, drinks and feeds going into human food chain. We will need to work with regulators to simplify and fast track certifications for algae foods.

27. Microcrops are simply too small to make a difference.

This statement from a USDA executive shows most farmers and farm policy leaders are unaware of microcrops. Farmers will benefit from the millions of wild microcrop species that produce 40% of each day's new biomass. Each alga cell can produce over a million offspring a day. Growers can harvest half the biomass daily, all through the growing season in open microfarms or all year in closed systems.

These tiny plants produce many times more protein per year than land crops without using cropland or freshwater. Small is beautiful in nature as microcrops offer higher nutralence and are more easily digested than food grains. These tiny plants also deliver superior nutrition to animals and field crops. Hence, they reduce input costs for industrial farmers while improving food crop yields and quality.

Growers will be attracted to abundance for the larger yields of tastier and more nutritious produce for both algaculture and field crops. Gardeners and farmers are continually experimenting with new growing methods and plant varieties. Growers want to maximize productivity while practicing predominantly organic gardening. They also want to minimize input costs and the use of agricultural chemicals. Produce grown with abundance methods offers ecobalance and health differentiation superior to organic.

Consumers are curious about microcrops but have little opportunity to try them because there are no commercial producers yet. Consumers want to minimize their ecological footprint, possibly by following the ecobalance diet. Mothers want to feed their children foods that will improve their brains, eyes, reading skills and behavior.[135] Families would like to serve foods without poison residues that cause allergic reactions, dull brains and developmental disabilities. Families will prefer serving foods that fight obesity and diabetes instead of causing those and other "Western" diseases.

Older people prefer to choose foods that repair their brains and prevent cognitive decline and dementia.[136] These benefits accrue to consumers of freedom foods with high nutralence, including omega-3 fatty acids and other antioxidants.

We also need to engage scientists, technologists and students with biotechnology knowledge. We need green masterminds with a passion for improving food quality and security for their family and community. We need educators who codify the training materials and convey abundance methods to people without reading or language skills. We need social, political, business and religious leaders who develop vision, values and action plans to move abundance forward. We need a green environmental groups and NGOs to promote the abundance environmental and health benefits.

We need media people who clearly articulate both the value proposition and the urgency for action. We need University professors and scientists to conduct the research that critically examines each of the value propositions for abundance. Most of all, we need our children to engage in this new food production model because their survival may depend on abundance methods and freedom foods. We encourage critiques; additions or insights at the open source web site AlgaeCompetition.com.

Chapter 12. Peace Microfarms Path Forward

For all the anxiety about the scarcity of oil, gas and vital minerals, the fiercest fight in the coming decades will involve food and the land it grows on. **– Michael T. Klare,** *The Race for what's Left: The Global scramble for the World's Last Resources, 2012.*

Michael Klare reports that resource depletion is accelerating faster than expected. He chronicles intense global competition for scarce resources and the frightening potential for war and environmental catastrophe. All countries will suffer from climate chaos, depleted cropland, scarce water and polluted ecosystems. We stand at a crossroads in food production with two paths. We can choose to:

1. Continue down the same path and produce primarily fossil foods relying on industrial agriculture, with full knowledge of the consequences for our children.

2. Develop peace microfarms to grow freedom foods with abundance methods that preserve natural resources and clean polluted environments.

Fortunately, we do not have to stop fossil food production. Farmers will continue using industrial methods to produce food – until the first of the 24 critical fossil resources becomes unaffordable. Expanding human populations will need every conceivable food source to supply sufficient food for the nine billion hungry mouths expected to share the earth's limited resources by 2050.

Climate chaos will create havoc on the food supply and push millions more farmers off their land. Farmers will deplete the modest fossil aquifer reserves and leave millions of cropland hectares abandoned with dry wells. Those once-fertile croplands will revert to dry prairie or desert. China, India and Africa provide living case studies for crashing aquifers and expanding deserts. The U.S. Mid-West and West will follow suit within the next generation.

People will have to migrate away from rural towns and cities when aquifers crash. Households need access to fresh water. Salt invasion from irrigation and sea surges will destroy additional cropland. No practical remediation exists for soil that becomes too saline to grow crops. Salt invasion, often accompanied by drought, has destroyed numerous prior civilizations.

Salt invasion ruins well water too. Tidal and storm surges often invade groundwater aquifers two or three times further than the surface water surge. Rising oceans will force millions from coastal cities and ruin prime coastal and delta croplands. Where will these people go?

Industrial agriculture's waste and pollution will continue to degrade and poison air, water and soil. Agricultural chemical and poison residue on foods will continue to cause children and adults to suffer illnesses and development disorders. Hidden hunger and empty calories will amplify the obesity and diabetes epidemic, creating severe drag for families, education and healthcare.

Novel solutions

Peace microfarms using abundance methods can avoid these predictable outcomes. Freedom foods are climate independent and expand the discussion about sustainable and affordable food beyond organic production. Food policy strategy needs to include microcrops at the base of the food chain. Microcrops can be grown with plentiful resources that do not compete with industrial foods. Many microcrops are salt tolerant so they flourish in water not useable by field crops, including wastewater, brine or ocean.

Microcrops have been overlooked in the future of food debates because no one offered a practical plan to overcome the joint challenges of commercial production and consumer acceptance. A set of talking points in Appendix I summarizes the arguments for peace microfarms. The early examples presented here provide encouraging evidence but prove neither commercialization nor enthusiastic buyer behavior. We have only just begun the fight to find a sustainable solution that augments our food supply.

Critical tasks

Peace microfarms will not magically appear because they seem to be a good idea. Serious work remains to convince growers to produce these crops and consumers to buy them. The first step is always the hardest – build new technologies that attract distributed growers.

Abundance methods are technology neutral in the sense that effective production will require each grower's thoughtful application of farming methods that are appropriate locally. Farmers consider a broad set of environmental conditions as well as input availability and costs. Growers also monitor markets needs and buyer behavior.

While abundance enables farmers to transition to predominantly organic methods, some may augment production with selected tools and techniques used in modern agriculture. For example, preliminary experience indicates growers may need to supplement one or more fertilizers because waste streams may not hold a full supply of replacement nutrients. Innovation research shows that incremental change occurs more successfully than a radical change such as a swift move from industrial to abundance farming.

Distributed small and medium size microfarms offers the highest potential for local jobs and social justice. Large farms concentrate wealth in the hands of a few. Large agribusinesses limit innovation to the relatively few people engaged in the firm. Large firms create value for their stakeholders by hoarding their intellectual property, which serves the company but not society.

Farmers and gardeners are clever and innovative. They will adopt abundance methods when the production models are sufficiently developed and field-tested. Every grower wants freedom from weather, water and waste. Growers want to improve productivity, reduce costly fossil resources and eliminate waste and pollution.

R3D

Microfarms need R3D, research, development, demonstration and diffusion. Research and development must design and build not just the growing systems but even more critical, grower support systems. Demonstration sites must prove that microcrops consistently produce yields that are a multiple of field crops, independent of weather or water. Growers must be able to produce superior food consistently with their local materials and inputs, in their climate and geography.

Peace microfarm diffusion to the many areas of need must wait until after the comprehensive research plan demonstrates growers have consistent success. Successful diffusion depends on positive grower experiences, and early adopter results will be critical.

Roll out will employ a two-step flow strategy where Green Masterminds, initial growers, receive considerable training and support for demonstration microfarms. Green Masterminds then will train their neighbors and community, which will multiply growers in every region. Peer training provides highly credible content for new growers and has been common in agriculture for centuries. Remote monitoring will be critical to successful microfarm adoption.

Remote monitoring

R&D will beta test peace microfarms distributed in various locations. Growers will be supported with monitoring systems that use the Internet to send data to a lab staffed with algae experts, AlgaeCentral.org. The algae metrics system:

- Provides data on the culture health and vitality.
- Identifies when the culture needs addition inputs.
- Sounds an alarm if predators or weed algae threaten the culture.
- Guides growers with regular action recommendations.
- Provides the key metrics to validate productivity.

Microfarm monitoring and metrics will not only provide training and a safety net for growers but they will also provide transparency to those who may be skeptical of this new form of food production. Monitoring systems can build grower intelligence and a knowledge base that benefits all future growers. No practical knowledge base for non-scientists exists today for algae cultivation.

Algae metrics will monitor key culture parameters such as pH, temperature, culture density, nutrient availability and salinity. Growers will report harvest metrics in order to track production and microfarm productivity.

Algae metrics will assess the presence of grazers. Competing microorganism predators are normally present, but should remain sparse in healthy cultures. Green Masterminds will have an iPhone microscope picture app that allows them to take and send pictures of their culture in order to identify unusual microorganisms.

Weed algae also need to be monitored because, like terrestrial gardens, weeds compete for nutrients and diminish the percentage of the target compounds in the biomass. Weed algae are typically present but grower actions can keep the weeds from significantly diminishing production. In extreme contamination cases, the only defense for the grower is to harvest the algae, clean the microfarm, and start a fresh batch.

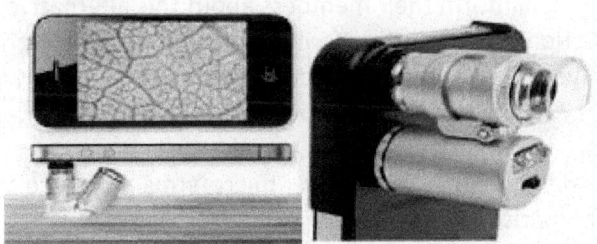

iPhone with attachable camera microscope

Growers and research scientists will want to know a wide set of production parameters such as growth rates and harvest data. Each grower will need to monitor component analysis to understand what fraction of the biomass is lipids, protein, carbohydrate or other compound. Specialty producers, such as nutraceuticals will want to measure what fraction of the lipids are omega-3 or other antioxidants. Omega-3 producers may need further analysis to monitor the presence of each EPA, DHA and other target compounds.

Demonstration

Demonstration sites create the opportunity to share peace microfarm metrics in real time with environmental groups like the Sierra Club, Green Peace, World Wildlife Fund, The Heifer Project and the Environmental Working Group. Environmental groups need access to production metrics to validate the ability to grow clean food from sterilized waste streams. The United Nations and FAO need similar access to peace microfarm production data so report summaries can be integrated into food policy documents.

Environmental groups that focus on specific fossil resources can use peace microfarm production data to convey this new form of resource preservation. The Pacific Institute, led by Peter Gleick, publishes the annual *The World's Water*, but does not yet recognize abundance methods as a way to preserve scarce water resources.[137] The Sustainable Phosphorus Initiative, in the School of Sustainability at Arizona State University promotes algae for nutrient recovery but has very little field data on nutrient recovery or microfarms.[138]

Professional horticulture, agriculture, aquaponics and phycology organizations can inform their members about this alternative form of food production. Peace microfarms sited at strong agricultural research universities such as UC Davis, Cornell, Texas A&M and the University of Arizona can create robust production data for scientific publications. Universities and research institutes can analyze production data from distributed microfarms for ecology and economic comparisons with modern agriculture.

Siting peace microfarms in high traffic locations such as zoos, botanical gardens, science centers and aquariums will convey the microfarm value proposition to millions. ZooPoo can transform a huge cost, waste management, into a profit center for the zoo while creating a superb eco-exhibit.[139] These sites can engage guests, students and volunteers in research projects as well as provide valuable production metrics. Public sites also offer extraordinary opportunity for targeted innovation on vital local issues such as sustainability, urban gardens and local jobs.

Special exhibitions, such as the Smithsonian and the Sustainable Food Exhibition in Milan, Italy in 2015 can demonstrate the value of peace microfarms to millions of visitors. The AlgaeFuture.org site offers excellent content for sustainable living and the future of food exhibits that may be located anywhere.

Microfarms can demonstrate urban abundance as growers produce food and feeds locally. Urban gardeners can recycle waste stream nutrients and clean water as they grow algae biofertilizers. Engaging Master Gardeners with smartcultures will validate increases in crop yield and quality while decreasing costs and waste.

Microfarms operating in government research locations such as the Cooperative Extension Service farms will demonstrate how these systems support field crops, reduce farming costs and minimize or eliminate waste streams. The Cooperative Extension Service can provide excellent test beds and innovation opportunities. Cooperative Extension agents understand horticulture and communicate with credibility to farmers.

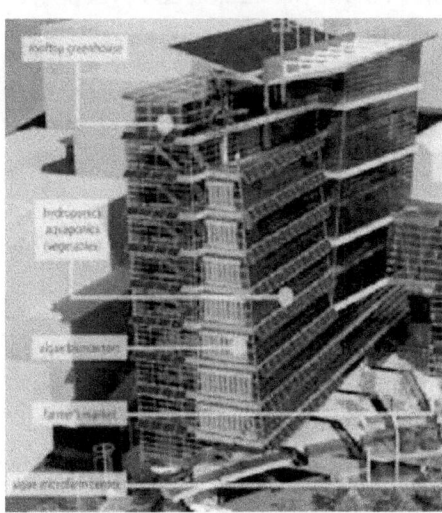

Designed by Mark Buehrer, CEO of 2020 Engineering, features:

- Rooftop greenhouse
- Hydroponics for vegetables
- Algae microfarms
- Farmer's market
- Algae microfarm center
- Soil-based crops fed with algae biofertilizer
- The EcoLab Algae Microfarm uses net-zero energy and produces net-zero waste.

Living Building

Entrepreneurial opportunities

Entrepreneurs need to develop and demonstrate effective microfarms that allow diverse growers to produce good food and a multitude of coproducts. Microfarms need to be affordable growing systems usable by nearly anyone, practically anywhere. Growers have used open cultivation systems for decades, but not very efficiently.[140] Current growing systems are too expensive and require far too much expertise and labor to operate.

Microfarms will need comprehensive yet fast training for Green Masterminds. The training materials will include videos and tutorials demonstrating every step in the growing process to minimize the need for grower education and reading. Diffusion to the developing world requires that peace microfarms be very simple to operate.

Similarly, the monitoring and horticulture support system needs to be simple, yet assure consistent success. A robust support system will significantly reduce training time because Green Masterminds will receive advice at each stage of the growth cycle. Only a fast and easy to learn support system will motivate wide spread adoption.

Engineers know how to build microfarms but the capital cost, (Capex) needs to be reduced by a factor of ten. Current cultivation systems typically run above $10,000 and require considerable expertise. We need to create a suite of microfarms that start around $500 and scale to larger and more sophisticated systems to enable fast adoption.

Similarly, operational costs, (Opex), must be reduced and made easier, faster and better. Every operational step must be simplified. Please engage in our fascinating global collaboratory, AlgaeFuture.org with your ideas for our abundant food future.

Quality control systems need to be transparent to insure that no producers in the distributed production network grow contaminated foods, feeds or other products. Extended food labeling will be important to assure consumers that the products produced are healthier for people, producers and our planet.

Policy

Policy leaders will need to navigate the complex government regulatory environments, including the FDA, USDA, OSHA and EPA. Industrial foods that are known to cause health problems are grandfathered in the food supply system. New foods like freedom foods that prevent health problems could be held up for years with very expensive bureaucratic regulatory hurtles.

Microfarmers will benefit from inclusion in new food legislation for small producers. Homemade Food Acts have been passed in 32 states and allow home-based entrepreneurs to sell up to $50,000 annually in "non-potentially hazardous" foods such as breads, jams and candy. Cottage food laws legalize practices such as sales at farmers' markets that are already commonplace in many communities but under the FDA radar. Cottage food acts enable entrepreneurs to test the market and then upgrade to commercial facilities and regulations when they expand production. Of course, these food acts will have to be amended to include algae-based foods.

Microfarm diffusion

Non-governmental organizations, professional groups and individuals will be helpful in diffusing microfarms to growers. Several groups including Antenna have built algae microfarms in Africa.[141] NGOs provide valuable help with funding, construction and training. The groups also give vital administrative and cultural support.

Antenna Technologies Circular Tank Spirulina System

Service organizations including Rotary, Kiwanis and Engineers without Boarders have indicated a desire to assist with the diffusion of microfarms. Foundations such as Kellogg, Gates and Ford offer opportunities for both diffusion and applied research. Church service and development organizations may use microfarms to improve people's lives in countries where they have missions.

The Heifer Project may add microfarms to their list of transformative opportunities for needy families. Microfarms align with the Heifer Project mission to work with communities to end hunger and poverty and care for the Earth. Microfarms can provide high nutralence foods for people, their animals and their gardens.

Many experts suggest foreign aid should give food to needy nations. Gifting food is unsustainable and creates dependence. Wealthy countries can and have gifted food in the short-term but soon will have insufficient food to give. No nation has the underlying fossil resources to sustain large food gifts. In the near future, no nation will have sufficient wealth to provide food transportation. The FAO reports that food transportation costs are often 400% higher than transferring the technology to grow food locally.

Rather than gifting food, why not send peace microfarms? Microfarms offer a far cheaper way to provide aid to hungry countries than transporting food. Peace microfarms may revolutionize foreign aid.

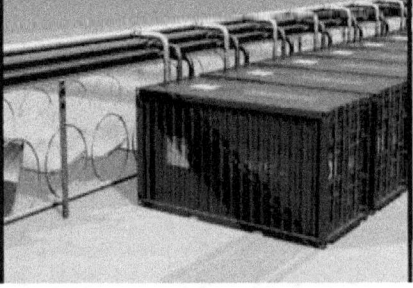

Open and Closed Microfarms deployable for Disaster Relief

Disaster relief offers another excellent opportunity. When disaster strikes, local people are often left with bad weather, destroyed crops and infrastructure, lots of botanical waste and scarce freshwater.

Microfarms can grow food quickly while remediating waste streams. In some disaster situations, the value of clean water may be higher than the food produced.

Microfarms offer a novel way of providing food for food banks. Imagine a food bank growing half the bank's food locally. Schools, universities, botanical gardens, aquariums and science enters could use microfarms as learning and training tools while producing good food, nutrients, feed or fertilizer. Homeless shelters could employ people in microfarm production and produce food for their neighbors.

Prisons could change their waste stream cost to a profit center while training inmates for green jobs. America is the only country in the world with over twice as many prisoners as farmers. Prisons offer a great opportunity to educate green values while producing food and cleaning waste streams. Special need farms have proven that food cultivation offers significant therapeutic value for a wide range of mental disorders commonly found among prisoners. The therapeutic value of freedom foods in prisons and special needs farms may exceed the economic value for food and waste management.

Wounded warriors

Peace microfarms would offer excellent entrepreneurial opportunities for wounded warriors and other veterans. The Wounded Warrior Program helps injured veterans build healthy and economically sound lives. With adaptable designs that fit specific capabilities, wounded warriors would be able to grow microcrops as an avocation or as a business venture. A microfarm might produce sufficient food for 75% of a family's needs. The same microfarm could produce the essential nutrient supply from a tablespoon of algae a day per person for 30 families. Entrepreneurial creativity will certainly discover many other production strategies.

The geographic siting flexibility and weather independence will enable veterans to grow microcrops for the needs of their family and community nearly anywhere. The wide range of target compounds derived from algae offer many business opportunities. Local needs are likely to drive production to those products or ecological pollution solutions most needed locally.

Many wounded warriors and others with disabilities could operate a microfarm adapted to their abilities. Young people join the military with a desire to serve their country. Microfarms give growers the opportunity to produce tangible products that offer value to their family and community.

Peace microfarm diffusion to wounded warriors might use a creative combination of subsidies and low-interest loans.

- Microfarm cost and installation = $10,000 one time.
- Operational costs = $5,000 a year.
- Revenue = $50,000 a year.

The first beta-test set of ten microfarms might be subsidized to demonstrate efficacy and refine operational issues. Additional microfarms might be subsidized at $5,000 each with a low-interest loan for the additional $5,000, plus the first year's operational costs. Microfarmers could then pay back their initial loans from microfarm revenues over a three-year period. Other options include models the solar industry and from successful microloans for other types of business.

Peace microfarms targeted to specific groups offer social and economic solutions for nearly any disadvantaged group such as Indian Reservations, areas of rural or urban poverty, single mothers, the elderly and ethnic groups suffering high unemployment and poverty.

Algae and national security

Algae and peace microfarms can play a major role in national security.

Military. Algae can remove the two primary reasons most wars are fought by providing sustainable and affordable food and avoiding the use of finite natural resources in food production.

Military operations. Peace microfarms at forward operating posts could save the military millions of dollars in logistical supplies, e.g. food, nutrients and fresh water, and save many soldier deaths in transportation.

Military transportation fuel. Large microfarms could produce fuel on military bases and avoid long and vulnerable supply routes.

Food security. Peace microfarms can provide food security for communities by recycling sanitized waste streams to supplement the food supply.

Water security. In some locations, algae's ability to create clean water will have more value than the biomass and coproducts.

Homeland security. Rather than building fences to block people's access to America, why not distribute peace microfarms to Mexican farmers to give them a way to make a living. Giving demonstration models and providing training, remote culture monitoring and coaching to growers would create a vital new industry and build a sustainable food model for the world. The entire peace microfarm project would cost less than several miles of ugly American boarder fence. Peace microfarms would offer a far more positive positioning for homeland security than a wall.

Childhood hunger security. One in five American children live in homes with food insecurity. Peace microfarms can provide access to healthy foods and end hidden hunger and nutrient deficiencies. Microfarmers local to consumers can improve food access while reducing the cost of food.

Elderly nutritional security. Today, a majority of Americans 65 and older do not get sufficient nutrition. Microfarms can give the elderly nutrition options from algae supplements or freedom foods..

Health security. Microfarms can grow nutrients, vaccines and medicines that end nutrient deficiencies and improve health and vitality. Microfarms offer freedom foods that fight the many diseases caused by inflammation as well as other "Western diseases."

Obesity and diabetes security. Freedom foods offer a natural solution to obesity and diabetes by providing high nutralent foods that are low in fat, cholesterol and have no empty calories. These algae-based foods manage insulin release and diminish the "nosh response" and replace it with a contented feeling of satiety.

Algae offer a wide range of solutions valuable for national security. The only national security funding provided by the U.S. government for algae has been for the development of algae biofuels. All the other applications have great value but remain, to date, largely unexplored.

Peace microfarm goals

Peace microfarms offer numerous sustainable outcomes besides avoidance of war and independence from weather, water and waste. Possibly the most significant are health, social justice, new jobs and transportation fuel savings. Radically reducing obesity and diabetes may be a high priority for the many countries afflicted. Resolving obesity related diseases could save tens of billions in health care and educational costs and restore positive lifestyle for millions of families.

Freedom foods may not fully resolve some target diseases. However, current cognitive and medical approaches have not slowed the increase in diabetes or several other Western diseases. A novel healthy foods model without empty calories offers a new natural solution that warrants medical, scientific and public debate.

Transforming food production to a distributed model with many producers close to consumers will improve social justice and access to fresh good food. People who currently cannot grow their own food due to climate, available cropland or economics may have a stronger opportunity to produce for their family with microfarms.

Local production will create tens of thousands of new green jobs. Growers will produce fresh natural foods for their family, neighbors and farmers markets. High productivity will allow growers to produce large amounts of food in relatively small space. In addition, microfarms can be sited on land or buildings that have no or minimal alternative use, which reduces site costs.

Aquamole producers and consumers in France

Enhance health and vitality

- Stop hidden hunger.
- End pesticide residues.
- Moderate obesity and diabetes.
- Resolve nutrient and vitamin deficiencies.
- End genetically engineered monocultures.

Elevate nutritious food

- Improve color, texture and taste.
- Grow nutraceuticals in the foods.
- Grow diverse energetic foods anywhere.
- Increase nutralence – nutrient availability and density.

Social

- Moderate poverty and hunger.
- Improve social equity; access to food.
- Enable food security for people everywhere.
- Moderate malnutrition with local healthy foods.

Environment

- Create climate independent foods.
- Reduce ecological pollution by 90%.
- Reduce freshwater consumption 90%.
- Reduce natural resource consumption 80%.

Food production economics

- Reduce total costs by 30 to 50%.
- Increase food productivity > 30 times.
- Cut transportation costs by > 50%.

Local foods

- Enable local production.
- Develop self-sufficient cities.
- Design consistent year round production.
- Grow foods in locations with no alternative use.

Table 12.1 *Peace Microfarm Goals*

Some green masterminds will sell their food products fresh, for direct consumption. Fresh foods may be made into dips or sauces such as aquamole, algaecream or algae ice cream. Others will sell alfu, which can substitute for tofu for use in soups, stews and salads. Algae flour products will include anything made with food grains, including bread, tortillas, chips, cakes and pies.

Food transportation cost reduction offers a valuable benefit set. The 50/50 freedom foods model with 50% of the food grown within 50 miles of consumers would save five times more liquid transportation fuel annually than the entire ethanol program. It would also save the 50 million acres of prime cropland planted in corn ethanol feedstock. Removing 50% of food trucks from our highways will substantially reduce pollution, road maintenance, and trucks that crush cars and people. A substantial reduction in black soot particulates in cities will reduce congestion, smog, and respiratory diseases significantly.

Local production reduces the over 50% of food costs required in packaging, preservatives, storage, transport, shrinkage and spoilage. Consumers will benefit from healthier, fresh local foods. Unlike the medical solutions that require further R&D, the local freedom foods production model is not theoretical but practical. Local food production saves trillions of gallons of fuel, trillions of gallons of water, millions of cropland acres while preserving natural resources.

The sustainability goals will need public policy support in order to change from the tyranny of the present. Vested interests, big agribusiness, could take substantial losses in a new food production model. Only a few people monopolize farm policy today, while many people are victims of their reach for profits. Fortunately, this new food model will work best by engaging small, medium and large farms to work jointly for food production and resource preservation.

Path forward

Our children's future depends on food security. Three critical threats to food security are global climate chaos, water scarcity and ecological pollution from agricultural waste. The current food supply will be decimated by more fierce storms, hotter and more acidic oceans, rising sea levels, higher temperatures, hot dry winds and prolonged drought. The availability of affordable fresh water has already passed the tipping point in many good growing areas. Additional heat and drought will accelerate water loss and prove fatal for many farms, including in the U.S. Midwest and West.

Countries will go to war over the natural resources vital to food production unless farmers and food policy leaders find ways to preserve fossil resources. Access to sufficient water for food production threatens nearly every country in the Mid-East today. Developing a food supply independent of fresh water and that cleans polluted water makes strategic sense.

Many people cannot grow food today due to weather or the cost of crop inputs. Creating a climate independent food supply will allow growers to produce fresh food year round. Designing food production to recover and repurpose waste stream energy and nutrients will assure growers can afford crop inputs. Microfarms also can moderate the waste and pollution caused by modern agriculture.

Green masterminds can grow high nutralence foods that end the nutrient deficiencies that severely limit growth and development of their children. Microfarms will help to clean ecosystems, which will improve the health and vitality of families that currently must live in waste stream plumes.

Farmers will embrace the opportunity to convert the costs associated with their waste stream to a profit center with the recovery and reuse of energy, nutrients and water. Society will celebrate the process of recovering energy and nutrients from waste streams while reversing air, water and soil pollution.

Urban growers will adopt abundance methods to produce freedom foods in their neighborhood. Developers will integrate microfarms in vertical farms, rooftop gardens and integrated living and growing spaces. Public architecture will become greenscapes with microfarms designed to provide visual, health and nutritional benefits.

Public policy discussions will focus on net-zero use of fossil resources and net-zero waste. Microfarms that remediate waste streams will be designed into living buildings that serve multiple purposes while creating a net-zero ecological footprint. Consumers will look for fresh local foods with higher nutralence and no empty calories. Local markets will sell foods with superior nutrition and taste that are grown with no pollution or waste.

Vertical farm   Rooftop garden

New industries will emerge to facilitate the engaging R3D that remain for peace microfarms. Entrepreneurs will develop new microfarm applications, new crops and new compounds such as medicines, nanomaterials and green chemicals. Green masterminds will find new pollution solutions that fit their communities. Growers will discover novel crops and diverse cultures that produce far more food than current methods. Smartcultures users will find more efficient ways for recovering farm or garden waste stream nutrients and delivering those nutrients to improve the yield and quality of field produce.

Peace microfarms using abundance methods will not resolve all food challenges. Unanticipated issues are certain to arise in abundance production, which is why transparent and accessible monitoring systems are needed. Microscope camera apps on mobile phones along with Internet connections make monitoring systems possible and practical.

Nature has used algae to support terrestrial plants for over 500 million years – since land plants evolved from algae. Now is the time that we need to mimic nature and enable algae to support sustainable, largely organic food production in a manner that uses fewer external inputs, lowers costs and reduces pollution.

Algae are available all over the Earth and are prepared to do their miraculous work to support growers, hungry consumers and society. We must act now to take advantage of algae's green promise – to create food security for everyone on our planet.

The design of affordable and easy to use microfarms remains our primary challenge. Collaborative open source social networks such as www.AlgaeFuture.org can provide practical solutions. We can pool our knowledge and share it with those who will transform our food supply system – green masterminds.

Our collective actions will enable people globally to grow sustainable and affordable food and valuable coproducts for their family and community locally.

President John F. Kennedy said:

> *We have the ability, we have the means, and we have the capacity to eliminate hunger from the face of the earth. We need only the will.*

Algae give us the means to eliminate hunger.

Our small team at AlgaeCompetition.com has the will.

Will you join us?

Appendix I. Peace Microfarm Food System Design Constraints
The Food Security Strategy Debate – Talking Points

Challenge	Description
	Consumer Health
1. Food security	Gives all people have access to affordable good food.
2. Consumer health	Delivers superior foods that naturally maximize health and vitality.
3. Fights diseases	Fights inflammatory diseases and improve major organ function including brain, eyes, heart, skin and respiratory system.
4. Avoids obesity	Low fat and cholesterol foods provide high quality protein, lipids, carbohydrates and other energetic compounds. Algae create satiety, which diminishes the urge to nosh.
5. Essential nutrients	Provides all 28 essential nutrients including essential vitamins, antioxidants, minerals and trace elements.
6. Nutralence	Delivers foods with high nutrient availability and density. Avoids hidden hunger and empty calories.
7. Sensory appeal	Foods have excellent color, aroma, texture, taste and mouth appeal.
8. Pesticide free	Clean and free of toxins, pathogens, fertilizers and poison residues.
9. Essential fatty acids	Micronutrients include omega-3 fatty acids and antioxidants.
10. Fresh and local	Over 50% of food can be grown within 50 miles of the consumer.
11. Transportation	50% cut in food transportation costs.

12. Black soot	Truck traffic reduction reduces black soot.
13. Natural biodiversity	Natural biodiversity avoids GE seeds and genetic monocultures.
14. Carbon footprint	300% lower carbon footprint.
15. Ecological footprint	500% lower ecological footprint than industrial food.
Grower Health and Risk	
16. Grower health	Growers and their families are free from health risk from agricultural chemicals.
17. Grower safety	Growers are free from physical risk from heavy labor and industrial machinery.
18. Rural health	Growers, families, farm animals and neighbors are free from agricultural pollutants and poisons.
19. Grower risk	Growers and their families are relieved of production risk due to possible crop failure.
20. Dust and air pollution	Avoids dust and air pollution by not disturbing the soil.
21. Waste	Avoids nutrient and water waste by recycling the culture. The only thing given off by algae production is pure oxygen.
22. Productivity multiple	Grower consistently produces 20 to 30 times more food per unit of cultivation every year independent of weather.
23. Affordable costs	Input costs are affordable and not tied to the price of fossil resources like fuel.
24. Nutrient extraction	No nutrient extraction because crops are grown in water on non-croplands.
25. Sustainable production	Freedom foods offer sustainable production for at least 7 generations.

	Ecological health
26. Fertile soil	Grows in containers that can be placed on non-cropland.
27. Fresh water	Grows in brackish, brine, waste and ocean water and can avoid freshwater.
28. Fossil fuels	Uses little energy that can be delivered from renewable sources.
29. Fertilizers	Can recover nutrients from waste streams and avoid mined inorganic fertilizers.
30. Poisons	Grow without pesticides, herbicides and fungicides by using natural biopesticides and controlling culture parameters.
31. Air remediation	Cleans air of gasses and black soot.
32. Cleans water	Cleans polluted water.
33. Soil regeneration	Restores soil fertility.
	Sociological health
34. Social equity	Everyone has access to affordable good food or the inputs to produce food.
35. Nutrient deficiencies	FAO reports that 50% of children globally suffer from nutrient deficiencies. Freedom foods can end nutrient deficiencies.
36. Economic risk	Growers know their crops will produce and not fail. Crop inputs are affordable.
37. Political risk	Governments and corporations do not control of food distribution and access.
38. Community health	Growers can form cooperatives to produce vitamins, minerals, nutraceuticals, vaccines, pharmaceuticals and medicines to promote family and community health.

39. Local jobs	Distributed peace microfarms can grow food nearly anywhere; growers enjoy good jobs.
40. Food independence	Distribution of freedom food knowledge globally will enable people to grow good food and other products for their family and community locally.
Smartcultures Benefits – Algae Biofertilizer for field crops	
42. Yield	Increase yield 20 to 80%.
43. Size	Increase produce size 20%.
44. Germination rate	Increase germination rate 20%.
45. Growth rate	Increase growth rate 10%.
44. Maturity	Speed time to grow to maturity 10%.
45. Color	Improve color 20%.
46. Sensory value	Improve aroma, taste and texture 20%.
47. Self life	Increase self-life 30%.
48. Water consumption	Smartcultures cut field crop water consumption by 20 to 30%.
49. Fertilizer consumption	Reduce fertilizer requirement by 60% for N and 40%+ for P and K. Micronutrient costs go to zero through waste stream recovery.
50. Pesticide use	Smartcultures can decrease the need for pesticides by 50% and fungicides by 75%.
51. High pH	Algae biomass that grows in the field can reduce elevated soil pH to normal levels.
52. Compression	Smartcultures deliver algae biofertilizers that can increase soil porosity by 500%.
53. Salt invasion	Improves soil porosity and allows irrigation salts to percolate below the root zone.

54. Humus loss	Rich algae biomass grows in the field, adding organic matter and natural fertility.
55. Pest control	Biofertilizers and growth hormones enable plants to produce their own biopesticides.
56. Crop stress	Algae biofertilizers improve plant health and vitality. Crops are able to withstand weather stress as well as pest vectors.
57. Erosion	No erosion from wind or water because the soil is not disrupted or degraded.
58. Pollution	Everything is recycled and repurposed, leaving no waste to pollute.

Note: A 20% difference in an attribute represents a scientifically significant difference.

Acknowledgements

New ideas build on the considerable research provided by science and environmental prior pioneers, including:
- David and Marcia Pimentel, emeritus, Cornell University
- Lester Brown, President of the Earth Policy Institute
- Jeffery Sachs of the Earth Institute at Columbia University
- Fred Krupp, president, the Environmental Defense Fund
- Ken Cook, President of the Environmental Working Group

Professors Qiang Hu, Milton Sommerfeld and Bruce Rittman from Arizona State University supported questions on molecular biology and algae production.

Science	Business – Econ.	Agribusiness
• Al Darzins	• Dan Simon	• Jon Ewen
• Mike Siebert	• David Schwartz	• Richard Morrison
• Jim Lane	• Mike Pasqualetti	• Ben Cloud
• Dan Childers	• Mark Allen	• Gary Wood
• James Elser	• Alan Resnik	• Doug Young
• Carol Johnston	• Gary Dyer	• Jim Robertson
• Gary Dirks	• Herb Roskind	• Tracy Penwell
• Andy Ayers	• Chris Low	• Barry Spiker

Also helpful were the published works of Paul Ehrlich, Sandra Postel, Nobel Laureate Al Gore, Harvey Blatt, Fred Pearce, Michael Pollen, Brian Halweil, Clay Jason and Linda Graham. High-content websites were a great support such as Algaebase, U.N., W.H.O., the National Resources Defense Council, Sierra Club, Green Peace, Audubon Society, Union of Concerned Scientists, Center for Energy and Climate Solutions, Clean Water Network and Public Citizen. Also useful were government sources including DOE, EPA, U.S. DA, NOAA and NREL.

Mark Edwards

Mark pursues R3D for sustainable and affordable peace microfarms designed to prevent war by producing good food while preserving vital resources. Microfarms improve health with better nutrition, clean ecosystems and provide jobs for millions.

Mark graduated from the U.S. Naval Academy where he earned degrees in engineering, oceanography and meteorology. Jacques Cousteau motivated and mentored his interest in the oceans and global stewardship. He holds an MBA and PhD in marketing and consumer behavior and has taught food marketing, engineering, sustainability, agribusiness, leadership and entrepreneurship at Arizona State University for over 30 years.

Mark served as marketing director for the Longevity Research Institute directed by Nathan Pritikin. The LRI focused on actions designed to improve the diet and exercise behaviors for people with health needs. The work led to the Pritikin diet and to Pritikin health foods. He also served as a director for a Fortune 50 foods company and has done extensive R&D on new foods, sources and consumer behavior. He has consulted for Monsanto, DuPont, Nabisco, Quaker Oats, General Mills, Borden, Coca-Cola, Frito-Lay, Disney, GE, Intel, J&J, Merck, GM, Bank of America, and many other companies.

He has published over 120 articles and 21 books that span business and science disciplines. His *360° Feedback* was a business best seller. Several have won best science and environment book awards including *Green Algae Strategy, Abundance: Sustainable fossil-free Food, The Tiny Plant that saved our Planet* and *Freedom Foods*. Colleges in in over 30 countries use some of the Green Algae Strategy series in food, energy and sustainability courses. Mark cofounded the global AlgaeCompetition with Robert Henrikson to enable people globally to produce food and coproducts for their family and community locally.   drmetrics@gmail.com

Robert Henrikson

Robert Henrikson has been a green business entrepreneur for over 30 years in sustainable development business models for algae, bamboo and natural resources. Robert was a founder of Earthrise Farms and for 20 years, was President of Earthrise, a pioneer in algae.

Robert was the creator and director of the International Bamboo Building Design Competition (BambooCompetition.com), and the former CEO of a leading company building certified, code-approved bamboo buildings. Robert is the co-author of the book *Bamboo Architecture in Competition and Exhibition*.

Robert serves as an AlgaeAlliance.com consultant on business development, strategic planning, branding, sales and marketing, advising companies and investors in algae ventures. He developed Earthrise® brand products in the USA and 30 countries. Authored the book *Spirulina World Food* in 2010, previously *Earth Food Spirulina*, translated into 6 international editions (SpirulinaSource.com).

Robert has written numerous articles and produced many videos on algae over the past 30 years, and currently contributes articles to Algae Industry Magazine and speaks at algae conferences. In 2011 he launched the International Algae Competition: A Global Challenge to Design Visionary Algae Food and Energy Systems.

Robert is a photographer (Panmagic.com) and documentary filmmaker, and produced the DVD series *Folding Time and Space at Burning Man* (Folding-Time.com). He is co-owner of Hana Gardenland, a botanical paradise retreat in Hana Maui, with vacation retreats and eco-tourism (HanaPalmsRetreat.com and Wild Thyme Farm, a sustainable forestry and permaculture farming eco-community. Email: roberthe@sonic.net.

The Green Algae Strategy Series

Mark R. Edwards

The Green Algae Strategy Series focuses on creating Sustainable and Affordable Food and Energy – "SAFE" production. **The Green Algae Strategy Series** are available for free downloading in color PDF for students, teachers and food and energy policy leaders at www.AlgaeFuture.org. They are also available on Amazon.com and other retailers. Teachers, professors and policy leaders use these SAFE production books in schools and colleges globally for courses in sustainability, engineering, business, politics, social entrepreneurship, food, water, energy, ecology, environment and world future.

BioWar I: Why Battles Over Food and Fuel Lead to World Hunger, 2007. BioWar I, where food is burned for fuel, must be ended by withdrawal – not of soldiers, but of damaging agricultural subsidies.

Green Algae Strategy: Engineer Sustainable Food and Fuel. 2008. Algae offer solutions for sustainable and affordable food and energy because algae are the most productive biomass source on Earth. *Best Science Book* **– 2009, Independent Publisher Awards**.

Green Solar Gardens: Algae's Promise to End Hunger, 2009. Algaculture in small but beautiful solar gardens and algae microfarms distributed globally will enable SAFE production locally.

Crash: The Demise of Fossil Foods and the Rise of Abundance. 2010. Traditional fossil-based agriculture sits precariously on a foundation of unsustainable fossil resources that will become unaffordable and then will run out. Abundant agriculture is sustainable because it uses plentiful inputs that are cheap and will not run out.

Smartcultures: Nature's tiny Genius – Algae – Reverses Pollution and Regenerates Degraded Ecosystems, 2011. Farmers may recycle farm wastes to their fields using abundance microfarms. Smartcultures give 20 – 30% higher yields by providing bioavailable nutrients at just the right time. Farmers save 30 – 40% by reducing input costs and reduce ecological pollution by 90%.

Abundance: Sustainable Fossil-free Foods with superior Nutrition and Taste; less Pollution and Waste, 2010. Abundance presents the value proposition for algae food, feeds and other forms of energy using plentiful resources that will not run out. Abundance growers can clean the air and water while they grow foods with superior nutrient density and better sensory values, including color, texture and taste. **Pinnacle Gold Medal winner 2011, Best Environmental Book.**

The tiny Plant that saved our Planet. The incredible true story of Tiny, Mighty Al. Tiny Mighty Al saved our planet by eating the bad carbon genie, which enabled the earth to cool and gave us oxygen. Al saved us again by becoming the bottom of the food chain and providing all living creatures with nutritious food. If we educate our children, maybe they will prompt us to take action — now. **Nautilus Silver Medal winner 2011, Best Children's Book.**

Abundant Agriculture: Smartcultures enable superior Nutrition and Yields from Regenerated Fields, 2010. Abundant agriculture represents the first new form of agriculture in 60 years. Abundant agriculture produces sustainable food, feed, fiber and other coproducts using primarily non-fossil resources that are plentiful, affordable and often surplus. Abundant agriculture growers use abundance methods to produce healthy, nutritional foods.

Freedom Foods: Superior Nutrition and Taste from low on the Food Chain for People, Producers and Our Planet, 2011. Freedom foods liberate consumers to make healthier food choices. Freedom foods are sustainable and grow free of fossil resources, GMO material and agricultural chemicals and pesticides.

Imagine Our Algae Future: **Visionary Algae Architecture and Landscapes**, 2012, with Robert Henrikson. See visionary images from the AlgaeCompetition.com showing how algae will change our world. Contestants from 40 countries created amazing graphics, pictures and videos showing how algae is produced today and will be used tomorrow for food, feed, biofuels, medicines and ecological repair.

Index

References

[1] Reuters, September 22, 2009, Obama's U.N. speech.
http://www.reuters.com/article/idUSTRE58L2PR20090922.

[2] Klare, Michael T. 2001. *Resource Wars: the New Landscape of Global Conflict.* New York: Metropolitan Books.

[3] http://daraint.org

[4] Brown, Lester. *Full Planet, Empty Plates: The New Geopolitics of Food Scarcity*, Earth Institute, 2012.

[5] http://www.savethechildren.org/site/

[6] USDA, Food stamps make America stronger, 2010, www.fns.usda.gov/

[7] Morgan Stanley, Little minds need big meals, *WSJ*, Dec 15, 20011, A1.

[8] Feed America, http://feedingamerica.org/SiteFiles/child-economy/

[9] FAO, Meat and meat products, *Food Outlook* FAO, Rome, December, 2006.

[10] Steinfeld, H. and Shalonda, P. Old players new players, *Livestock report 2006*, Rome: FAO, 2006, 3.

[11] Steinfeld, H. *Livestock's long shadow*, Rome: FAO, 2006.

[12] http://www.askthemeatman.com/yield_on_beef_carcass.htm

[13] Falk, Dean, *Fresh Water Needs for a Dairy.*
http://www.oneplan.org/Stock/DairyWater.shtml

[14] Pearce, Fred. *When the Rivers Run Dry: Water--The Defining Crisis of the Twenty-first Century*, Beacon Press, 2007.

[15] Nierenberg, Danielle. *Rethinking the Global Meat Industry.* State of the World: 2006. Ed. The World Watch Institute. London: Norton, 2006: 24.

[16] http://money.cnn.com/2012/08/03/news/economy/drought-crop-insurance/index.htm

[17] Kaplan, Robert. The Coming Anarchy', *Atlantic Monthly*, 2:2, 1994, 44–76.

[18] Christian Aid, 2007. Human Tide: the Real Migration Crisis. Christian Aid, http://www.christianaid.org.uk/Images/human_tide3__ 23335.pdf.

[19] Homer-Dixon, Thomas, Terror in the Weather Forecast, *New York Times*, 24 April 2007. www.nytimes.com/2007/04/24/opinion/24homer-dixon.html?_r=1&oref=slogin.

[20] Schwartz, Peter & Doug Randall, 2003. *An Abrupt Climate Change Scenario and Its Implications for U.S. National Security.* Report prepared for the U.S. Department of Defense.

[21] CNA. *National Security and the Threat of Climate Change*, 2007, 1. Report from a panel of retired senior US military officers. Alexandria, VA: CNA Corporation, http://securityandclimate.cna.org/.

[22] Christian Parenti. *The Tropic of Chaos: Climate Change and the New Geography of Violence*, Nation Books, 2012, 12.

[23] David D. Zhang, et al. *Global climate change, war, and population decline in recent human history*, PNAS, 104:49, Dec 4, 2007, 19214-19219.

[24] David D. Zhang et al. *The causality analysis of climate change and large-scale human crisis*, PNAS 108:42, October 18, 2011, 17296-17301.

[25] Pimentel, David and Marsha Pimentel. *Food, energy and society*, third edition, New York: CRC Press, 2008, 27.

[26] World Watch. *Cropland Losses Threaten World Food Supplies*, July 27, 1996. http://www.worldwatch.org/cropland-losses/

[27] World Bank: *Land Resources Management*, lnweb18.worldbank.org/ESSD/ardext.nsf/ 11ByDocName/

[28] http://www.ewg.org/losingground/report/executive-summary

[29] Kolbert, Elizabeth. *Field Notes from a Catastrophe: Man Nature and Climate Change*. New York: Bloomsbury, 2006: 96-97.

[30] Zoellick, Robert. *High-Level Conference on World Food Security, Rome, World Bank*, June 4, 2008. No: 2008/349/EXC.

[31] http://www.triplepundit.com/pages/un-rome-conference-mobilisatio-003225.php

[32] Biofuels, *Biofuels Digest*, http://www.biofuelsdigest.com/blog2/2008/03/07/today-in-biofuels

[33] http://www.larouchepac.com/node/10736

[34] http://deltafarmpress.com/news/food-procudtion-0604/

[35] *Annual Energy Report*, (PDF). US Department of Energy (2006-07).

[36] World Bank, *World Development Report* 2008: Agriculture for Development, October, 2007. http://publications.worldbank.org/ecommerce/catalog/

[37] Nicholas Stern, *The Stern Review on the Economics of Climate Change*, (London: HM Treasury, 2006).

[38] Edwards, Mark. *Biowar I: Why Battles over Food and Fuel Lead to World Hunger*, Tempe, 2008, 117.

[39] Renewable Fuels Association, http://www.ethanolrfa.org/pages/statistics#A

[40] Lewis, Leo. Scientists warn of lack of vital phosphorus as biofuels raise demand, The Times, June 23, 2008.

[41] http://sustainablep.asu.edu

[42] Diouf, Jacques. *Turning the Tide Against Water Scarcity.* Food and Agriculture Organization of the United Nations, Mar. 2007. http://www.fao.org/english/dg/oped/index.html.

[43] Pimentel and Giampietro. Food, Land, Population and the U.S. Economy, Nov. 1994.

[44] Fred Pearce, "Asian Farmers Sucking the Continent Dry," New Scientist.com, 28 August 2004.

[45] India Times, http://www.indiatribune.com/index.php?option=com_content&view=article&id=5389:every-12-hours-one-farmer-commits-suicide-in-india&catid=106:magazine

[46] Edwards, Mark R. *Abundance: Sustainable Fossil-free Foods,* 2010.

[47] Edwards, Mark R. *Abundant Agriculture: Smartcultures enable superior Nutrition and Yields from Regenerated Fields* Tempe: CreateSpace, 2010.

[48] Goyal, SK. A profile of algal biofertilizer. in *Biotechnology of Biofertilizers,* Kannaiyan, S. Ed., Delhi: Narosa Publishing House, 2002, 250 – 258.

[49] Edwards, Mark R. *Green Algae Strategy: Engineer Sustainable Food and Fuel,* CreateSpace, 2008, 44.

[50] A list of algae collections is available at AlgaeCompetition.com.

[51] Edwards, Mark. *Green Solar Garden: Algae's Promise to End Hunger,* 2009.

[52] Edwards, Mark. *Freedom Foods: Superior Nutrition and Taste from low on the Food Chain for People, Producers and Our Planet,* 2011.

[53] Warner, Jennifer. CDC: Kids Lack Access to Healthy Food Choices, WebMD Health News , April 26, 2011.

[54] Center for Food Safety, http://www.centerforfoodsafety.org /2011/03/18/

[55] http://www.cdc.gov/healthyyouth/obesity/facts.htm

[56] http://www.cdc.gov/diabetes/pubs/pdf/ndfs_2011.pdf

[57] http://www.diabetes.org/advocate/resources/cost-of-diabetes.html

[58] http://www.ers.usda.gov/briefing/organic/Farmsector.htm

[59] Edwards, Mark R. *Abundant Agriculture,* 2010.

[60] Cui-Hua Qi, Min Chen, Jie, Soong, Bao-Shan Wang. Increase in aquaporin activity is involved in leaf succulence of the euhalophyte suaeda salsa, under salinity. *Plant Science*; 176;2, 200-205, Feb 2009.

[61] Nutralence is a new word that refers to plants that naturally store nutrients, delivering higher availability and nutrient density.

[62] Jeff Norrie, Seaweed Research. *Am. Fruit Grower*, 128;3, 48-50, Mar 2008.

[63] Luescher-mattli, M. Current Medical Chemistry-Anti-Inflammatory Agents, in *Ingenta-Connect*, 2, 2003, 219-225.

[64] MacArtain P, Gill CIR, Brooks M, Campbell R, Rowland IR. Nutritional value of edible seaweeds. *Nutrition Review*, 2007, 65:535-543.

[65] Spolaore P, Joannis-Cassan C, Duran E, Isambert A. Commercial applications of microalgae. 2006 *J Biosci Bioeng* 101:87-96.

[66] Garcia-Casal, et. al. 2007.

[67] Garcia-Casal MN, Pereira AC, Leets I, Ramirez J, Quiroga MF. High iron content and bioavailability in humans from four species of marine algae. *Journal of Nutrition,* 2007, 137:2691-2695.

[68] Kavaler, Lucy. *Green Magic: Algae Rediscovered*. Thomas Crowell, NY, 1983, 99-101.

[69] Dhargalkar, V. K. and X. N. Verlecar. Southern Ocean seaweeds: A resource for exploration in food and drugs. *Aquaculture*, 287(3/4): 2009, 229-242.

[70] Lisheng, L. et. al. Inhibitive effect and mechanism of polysaccharide of spirulina on transplanted tumor cells in mice. *Marine Sciences*, Qindao, China. N.5, 1991, 33-38.

[71] Luescher-mattli, M., Current Medical Chemistry-Anti-Inflammatory Agents, 2010, 2, 219-225.

[72] Palmquist RE. 2008. Apparent response to homotoxicology, salmon oil and blue-green algae in a single geriatric canine case of episodic mentation changes. JAHVMA 27 (1): April-June, 2010, 10-15.

[73] McCarty MF, Barroso-Aranda J, Contreras F. NADPH oxidase mediates glucolipotoxicity-induced beta cell dysfunction--clinical implications. Med Hypotheses.;74(3):March, 2010, 596-600.

[74] Lee EH, Park JE, Choi YJ, Huh KB, Kim WY. A randomized study to establish the effects of spirulina in type 2 diabetes mellitus patients. Nutr Res Pract. 2008 Winter; 2(4): 295-300.

[75] Muthuraman P, Senthilkumar R, Srikumar K. Alterations in beta-islets of Langerhans in alloxan-induced diabetic rats by marine Spirulina platensis. *J Enzyme Inhib Med Chem*. 2009 Dec;24(6):1253-6.

[76] Garbuzova-Davis, Svitlana. *The Open Tissue Engineering and Regenerative Medicine Journal*, 2011, 3, 36-41.

[77] Gupta S, Hrishikeshvan HJ, Sehajpal PK. Spirulina protects against rosiglitazone induced osteoporosis in insulin resistance rats. Diabetes Res Clin Pract. 2010 Jan;87(1):38-43.

[78] Shytle DR, Tan J, et al. Effects of blue-green algae extracts on the human adult stem cells in vitro. Med Sci Monit. 2010 Jan, 16(1): BR1-5.

[79] Fountain, Henry. Too Many Small Fish Are Caught, Report Says, New York Times, April 2, 2012.

[80] http://www.oceanconservationscience.org/foragefish/

[81] http://www.dietresearch.com/

[82] Edwards, Mark R. Abundance, 91.

[83] Milton K. The critical role played by animal source foods in human (Homo) evolution. J. Nutr. 2003, 133:3886–92S.

[84] Edwards, Mark R. Algae 101 Blog, Part 26: Did Algae's Great Taste Make us do it? Algae Industry Magazine, May, 2011.

[85] Kulshreshtha, Archana et. al. Spiralina in healthcare management, Current pharmaceutical biotechnology, 2008, nine, 400 – 405.

[86] MacArtain P, Gill CIR, Brooks M, Campbell R, Rowland IR. Nutritional value of edible seaweeds. Nutr Rev 2007, 65:535-543.

[87] Yamada Y, Miyoshi T, Tanada S, Imaki M. Digestibility and energy availability of Wakame seaweed, Jap J Hygiene, 2007, 46, 788-793.

[88] Edwards, Mark R. Abundance, 88.

[89] Yangthong M, et al. Antioxidant activities of four edible seaweeds from the southern coast of Thailand. Plant Foods Hum Nutr 2009, 64:218-223.

[90] Abad MJ, Bedoya LM, Bermejo P. 2008 Natural marine anti-inflammatory products. Mini Rev Med Chem 8,740-754.

[91] Edwards, Mark R. Food, energy and habitat for the 100 Year Starship, NASA, 100 Year Starship Symposium, Orlando, FL, 2011.

[92] Graham, Linda and Lee Wilcox. Algae. New Jersey: Prentice Hall, 2008: 8.

[93] Megasun.bch.umontreal.ca/protists/gallery.html algaebase.org/links/ utex.org; ccmp.bigelow.org; http://www.ccap.ac.uk; marine.csiro.au/microalgae; wdcm.nig.ac.jp/hpcc.html).

[94] Hu, Qiang. "Environmental Effects on Cell Composition." Handbook of Microalgal Culture: Biotechnology and Applied Phycology. Ed. Amos Richmond. Oxford, England: Blackwell Science, Ltd., 2004: 83-94.

[95] Edwards, Mark R. Green Solar Gardens, 12.

[96] Edwards, Mark R. Smartcultures: Nature's tiny Genius – Algae – Reverses Pollution and Regenerates Degraded Ecosystems, CreateSpace, 2010, 14.

[97] Perry, Ann. Algae: A Mean, Green Cleaning Machine, *Agricultural Research Magazine*, 58:5, May/June 2010.

[98] http://www.parkseed.com/gardening/PD/9261/

[99] www.cooksgarden.com

[100] Oswald, W.J. and C.G. Golueke, Biological transformation of solar energy, *Advances in. Applied Microbiology, 2, 1960,* 223–262.

[101] *Borowitzka, Michael A.* Culturing Microalgae in Outdoor Ponds, in Andersen, Robert A., Ed. *Algal culturing techniques*, Phycological Society of America, Elsevier Academic Press, 2005, 205 – 219.

[102] Ugwu CU, Aoyagi H, Uchiyama H. Photobioreactors for mass cultivation of algae. *Bioresource Technology*, 2008, 99(10): 4021-8.

[103] Florida Department of Environmental Protection, www.bioreactor.org/

[104] Edwards, Mark. Unpublished survey research, ASU, 2008.

[105] Behrens, Paul W. Photobioreactors and Fermentors: The Light and Dark Sides of Growing Algae, in Andersen, A., Ed. *Algal culturing techniques*, Phycological Society of America, Elsevier Academic Press, 2005, 189 -205.

[106] USDA, http://www.ers.usda.gov/publications/err-economic-research-report/err141.aspx

[107] http://www.theatlanticwire.com/.../2012/07/us...disaster-area.../5447...

[108] McGovern, *Ending Hunger.* amazon.com/Ending-Hunger-Now-George-McGovern/dp/0800637828

[109] http://www.americanprogress.org/issues/poverty/report/2011/10/05/10504/hunger-in-america/

[110] http://www.fns.usda.gov/snap/

[111] www.nytimes.com/2009/05/24/opinion/24kristof.html

[112] http://www.hungeractionnys.org/health6.htm

[113] http://hcsmfish.org/Consequences of Hunger.pdf

[114] mayoclinic.com/health/iron-deficiency-anemia/DS00323

[115] http://www.cdc.org

[116] http://www.mayoclinic.com/health/beta-carotene/NS_patient-betacarotene

[117] http://www.mayoclinic.com/health/antioxidants/MY01593

[118] http://nutrition.tufts.edu

[119] tm.mahidol.ac.th/seameo/2008_39_1/25-4159.pdf

[120] ers.usda.gov/data-products/food-desert-locator.aspx

[121] agritech.tnau.ac.in/agriculture/agri_min_nutri_def_symptoms

[122] nytimes.com/.../global-food-prices-on-the-rise-united-nations

[123] http://www.sustainabletable.org/issues/energy/

[124] Edwards, Mark R. *Smarticultures*, 2010.

[125] Vaidyanathan, Gayathri. Genetic Engineering No Match for Evolution of Weed Resistance, *Scientific American*, April 14, 2010.

[126] Tisdale, S. L. and W. L. Nelson. *Soil Fertility and Fertilizers*. 3rd ed. New York: Macmillan, 1975.

[127] Brady, N. C. *The Nature and Properties of Soils*. New York: Macmillan Publishing Co., 1974.

[128] Gliessman, Stephen R. Agroecology: The Ecology of Sustainable Food Systems, Second Edition, CRC Press; 2 ed., 2006.

[129] Plaster, E. J. *Soil Science and Management*. 3rd ed. Albany: Delmar Publishers, 1996.

[130] Edwards, Mark R. BioWar I, 84.

[131] Prince Charles, Future of Food Conference, Georgetown University May 4, 2011. http://washingtonpostlive.com/conferences/food

[132] DOE, National Algal Technology Roadmap, 2010. http://www1.eere.energy.gov/biomass/pdfs/algal_biofuels_roadmap.pdf

[133] washingtonpost.com/blogs/wonkblog/wp/2012/08/10/where-the-worlds-running-out-of-water-in-one-map/

[134] //blogs.scientificamerican.com/observations/2012/08/09/farmers-deplete-fossil-water-in-worlds-breadbaskets/

[135] Oxford University. http://dx.plos.org/10.1371/journal.pone.0043909

[136] Sydenham E, et al. Omega 3 fatty acid for the prevention of cognitive decline and dementia. Cochrane Database, 2012 Jun 13;6: CD005379.

[137] http://www.worldwater.org

[138] http://sustainablep.asu.edu

[139] Edwards, Mark. Abundance, 2011.

[140] Robert Henrikson's blog, Microfarms, *Algae Industry Magazine*.

[141] Henrikson, Robert and Mark Edwards, *Imagine Our Algae Future*, CreateSpace, 2012, 42.